R. T. Glazebrook

**James Clerk Maxwell and Modern Physics**

R. T. Glazebrook

**James Clerk Maxwell and Modern Physics**

ISBN/EAN: 9783743318069

Manufactured in Europe, USA, Canada, Australia, Japa

Cover: Foto ©berggeist007 / pixelio.de

Manufactured and distributed by brebook publishing software (www.brebook.com)

R. T. Glazebrook

**James Clerk Maxwell and Modern Physics**

*THE CENTURY SCIENCE SERIES*

EDITED BY SIR HENRY E. ROSCOE, D.C.L., LL.D., F.R.S.

# JAMES CLERK MAXWELL
# AND MODERN PHYSICS

# The Century Science Series.

EDITED BY
SIR HENRY E. ROSCOE, D.C.L., F.R.S., M.P.

**John Dalton and the Rise of Modern Chemistry.**
By Sir HENRY E. ROSCOE, F.R.S.

**Major Rennell, F.R.S., and the Rise of English Geography.**
By CLEMENTS R. MARKHAM, C.B., F.R.S., President of the Royal Geographical Society.

**Justus von Liebig: his Life and Work (1803–1873).**
By W. A. SHENSTONE, F.I.C., Lecturer on Chemistry in Clifton College.

**The Herschels and Modern Astronomy.**
By AGNES M. CLERKE, Author of "A Popular History of Astronomy during the 19th Century," &c.

**Charles Lyell and Modern Geology.**
By Rev. Professor T. G. BONNEY, F.R.S.

**James Clerk Maxwell and Modern Physics.**
By R. T. GLAZEBROOK, F.R.S., Fellow of Trinity College, Cambridge.

*In Preparation.*

**Michael Faraday: his Life and Work.**
By Professor SILVANUS P. THOMPSON, F.R.S.

**Humphry Davy.**
By T. E. THORPE, F.R.S., Principal Chemist of the Government Laboratories.

**Pasteur: his Life and Work.**
By M. ARMAND RUFFER, M.D., Director of the British Institute of Preventive Medicine.

**Charles Darwin and the Origin of Species.**
By EDWARD B. POULTON, M.A., F.R.S., Hope Professor of Zoology in the University of Oxford.

**Hermann von Helmholtz.**
By A. W. RÜCKER, F.R.S., Professor of Physics in the Royal College of Science, London.

CASSELL & COMPANY, LIMITED, *London; Paris & Melbourne.*

*(From a Photograph of the Picture by G. Lowes Dickinson, Esq., in the Hall of Trinity College, Cambridge.)*

*THE CENTURY SCIENCE SERIES*

# JAMES CLERK MAXWELL

## AND MODERN PHYSICS

BY

### R. T. GLAZEBROOK, F.R.S.

*Fellow of Trinity College, Cambridge
University Lecturer in Mathematics, and Assistant Director of the
Cavendish Laboratory*

———•••———

CASSELL AND COMPANY, LIMITED

*LONDON, PARIS & MELBOURNE*

1896

ALL RIGHTS RESERVED

# PREFACE.

THE task of giving some account of Maxwell's work —of describing the share that he has taken in the advance of Physical Science during the latter half of this nineteenth century—has proved no light labour. The problems which he attacked are of such magnitude and complexity, that the attempt to explain them and their importance, satisfactorily, without the aid of symbols, is almost foredoomed to failure. However, the attempt has been made, in the belief that there are many who, though they cannot follow the mathematical analysis of Maxwell's work, have sufficient general knowledge of physical ideas and principles to make an account of Maxwell and of the development of the truths that he discovered, subjects of intelligent interest.

Maxwell's life was written in 1882 by two of those who were most intimately connected with him, Professor Lewis Campbell and Dr. Garnett. Many of the biographical details of the earlier part of this book are taken from their work. My thanks are due to them and to their publishers, Messrs. Macmillan, for permission to use any of the letters which appear in their biography. I trust that my brief account may be sufficient to induce many to read Professor Campbell's "Life and Letters," with a view of learning more of the inner thoughts of one who has left so strong an imprint on all he undertook, and was so deeply loved by all who knew him.

R. T. G.

*Cambridge,*
*December,* 1895.

# CONTENTS.

|   |   | PAGE |
|---|---|---|
| CHAPTER I.—EARLY LIFE . | . | 9 |
| ,, II.—UNDERGRADUATE LIFE AT CAMBRIDGE . | . | 28 |
| ,, III.—EARLY RESEARCHES  PROFESSOR AT ABERDEEN | . | 38 |
| ,, IV.—PROFESSOR AT KING'S COLLEGE, LONDON—LIFE AT GLENLAIR . | . | 51 |
| ,, V.—CAMBRIDGE—PROFESSOR OF PHYSICS |  | 60 |
| ,, VI.—CAMBRIDGE—THE CAVENDISH LABORATORY . |  | 73 |
| ,, VII.—SCIENTIFIC WORK—COLOUR VISION |  | 93 |
| ,, VIII.—SCIENTIFIC WORK—MOLECULAR THEORY |  | 108 |
| ,, IX.—SCIENTIFIC WORK—ELECTRICAL THEORIES . |  | 148 |
| ,, X.—DEVELOPMENT OF MAXWELL'S THEORY | . | 202 |

# JAMES CLERK MAXWELL
## AND MODERN PHYSICS.

### CHAPTER I.

#### EARLY LIFE.

"ONE who has enriched the inheritance left by Newton and has consolidated the work of Faraday —one who impelled the mind of Cambridge to a fresh course of real investigation—has clearly earned his place in human memory." It was thus that Professor Lewis Campbell and Mr. Garnett began in 1882 their life of James Clerk Maxwell. The years which have passed, since that date, have all tended to strengthen the belief in the greatness of Maxwell's work and in the fertility of his genius, which has inspired the labours of those who, not in Cambridge only, but throughout the world, have aided in developing the seeds sown by him. My object in the following pages will be to give some very brief account of his life and writings, in a form which may, I hope, enable many to realise what Physical Science owes to one who was to me a most kind friend as well as a revered master.

The Clerks of Penicuik, from whom Clerk Maxwell was descended, were a distinguished family. Sir John Clerk, the great-great-grandfather of Clerk Maxwell, was a Baron of the Exchequer in Scotland from 1707 to 1755; he was also one of the Commissioners of the Union, and was in many ways an accomplished scholar. His second son George married a first cousin, Dorothea Maxwell, the heiress of Middlebie in Dumfriesshire, and took the name of Maxwell. By the death of his elder brother James in 1782 George Clerk Maxwell succeeded to the baronetcy and the property of Penicuik. Before this time he had become involved in mining and manufacturing speculations, and most of the Middlebie property had been sold to pay his debts.

The property of Sir George Clerk Maxwell descended in 1798 to his two grandsons, Sir George Clerk and Mr. John Clerk Maxwell. It had been arranged that the younger of the two was to take the remains of the Middlebie property and to assume with it the name of Maxwell. Sir George Clerk was member for Midlothian, and held office under Sir Robert Peel. John Clerk Maxwell was the father of James Clerk Maxwell, the subject of this sketch.*

John Clerk Maxwell lived with his widowed mother in Edinburgh until her death in 1824. He was a lawyer, and from time to time did some little business in the courts. At the same time he maintained an interest in scientific pursuits, especially those of a practical nature. Professor Campbell tells us of an endeavour to devise a bellows which would give a continuous draught of air. In 1831 he

* A full biographical account of the Clerk and Maxwell families is given in a note by Miss Isabella Clerk in the "Life of James Clerk Maxwell," and from this the above brief statement has been taken.

contributed to the *Edinburgh Medical and Philosophical Journal* a paper entitled "Outlines of a Plan for combining Machinery with the Manual Printing Press."

In 1826 John Clerk Maxwell married Miss Frances Cay, of North Charlton, Northumberland. For the first few years of their married life their home was in Edinburgh. The old estate of Middlebie had been greatly reduced in extent, and there was not a house on it in which the laird could live. However, soon after his marriage, John Clerk Maxwell purchased the adjoining property of Glenlair and built a mansion-house for himself and his wife. Mr. Maxwell superintended the building work. The actual working plans for some further additions made in 1843 were his handiwork. A garden was laid out and planted, and a dreary stony waste was converted into a pleasant home. For some years after he settled at Glenlair the house in Edinburgh was retained by Mr. Maxwell, and here, on June 13, 1831, was born his only son, James Clerk Maxwell. A daughter, born earlier, died in infancy. Glenlair, however, was his parents' home, and nearly all the reminiscences we have of his childhood are connected with it. The laird devoted himself to his estates and to the education of his son, taking, however, from time to time his full share in such county business as fell to him. Glenlair in 1830 was very much in the wilds; the journey from Edinburgh occupied two days. "Carriages in the modern sense were hardly known to the Vale of Urr. A sort of double gig with a hood was the best apology for a travelling coach, and the most active mode of locomotion was in a kind of rough dog-cart known in the family speech as a hurly."*

Mrs. Maxwell writes thus†, when the boy was nearly three years old, to her sister, Miss Jane Cay:—

"He is a very happy man, and has improved much since the weather got moderate. He has great work with doors, locks, keys, etc., and 'Show me how it doos' is never out of his mouth. He also investigates the hidden course of streams and bell-wires—the way the water gets from the pond through the wall and a pend or small bridge and down a drain into Water Orr, then past the smiddy and down to the sea, where Maggy's ships sail. As to the bells, they will not rust; he stands sentry in the kitchen and Mag runs through the house ringing them all by turns, or he rings and sends Bessy to see and shout to let him know; and he drags papa all over to show him the holes where the wires go through."

To discover "how it doos" was thus early his aim. His cousin, Mrs. Blackburn, tells us that throughout his childhood his constant question was, "What's the go of that? What does it do?" And if the answer were too vague or inconclusive, he would add, "But what's the *particular* go of that?"

Professor Campbell's most interesting account of these early years is illustrated by a number of sketches of episodes in his life. In one Maxwell is absorbed in watching the fiddler at a country dance; in another he is teaching his dog some tricks; in a third he is helping a smaller boy in his efforts to build a castle. Together with his cousin, Miss Wedderburn, he devised a number of figures for a

---

\* "Life of J. C. Maxwell," p. 26.
† "Life of J. C. Maxwell," p. 27.

toy known as a magic disc, which afterwards developed into the zoetrope or wheel of life, and in which, by means of an ingenious contrivance of mirrors, the impression of a continuous movement was produced.

This happy life went on until his mother's death in December, 1839; she died, at the age of forty-eight, of the painful disease to which her son afterwards succumbed. When James, being then eight years old, was told that she was now in heaven, he said: "Oh, I'm so glad! Now she'll have no more pain."

After this his aunt, Miss Jane Cay, took a mother's place. The problem of his education had to be faced, and the first attempts were not successful. A tutor had been engaged during Mrs. Maxwell's last illness, and he, it seems, tried to coerce Clerk Maxwell into learning; but such treatment failed, and in 1841, when ten years old, he began his school-life at the Edinburgh Academy.

School-life at first had its hardships. Maxwell's appearance, his first day at school, in Galloway homespun and square-toed shoes with buckles, was more than his fellows could stand. "Who made those shoes?" they asked *; and the reply they received was—

> "Div ye ken 'twas a man,
> And he lived in a house,
> In whilk was a mouse."

He returned to Heriot Row that afternoon, says Professor Campbell, "with his tunic in rags and

---

* "Life of J. C. Maxwell," p. 49.

wanting the skirt, his neat frill rumpled and torn—himself excessively amused by his experiences and showing not the slightest sign of irritation."

No. 31, Heriot Row, was the house of his widowed aunt, Mrs. Wedderburn, Mr. Maxwell's sister; and this, with occasional intervals when he was with Miss Cay, was his home for the next eight or nine years. Mr. Maxwell himself, during this period, spent much of his time in Edinburgh, living with his sister during most of the winter and returning to Glenlair for the spring and summer.

Much of what we know of Clerk Maxwell's life during this period comes from the letters which passed between him and his father. They tell us of the close intimacy and affection which existed between the two, of the boy's eager desire to please and amuse his father in the dull solitude of Glenlair, and his father's anxiety for his welfare and progress.

Professor Campbell was his schoolfellow, and records events of those years in which he shared, which bring clearly before us what Clerk Maxwell was like. Thus he writes * :—

"He came to know Swift and Dryden, and after a while Hobbes, and Butler's 'Hudibras.' Then, if his father was in Edinburgh, they walked together, especially on the Saturday half-holiday, and 'viewed' Leith Fort, or the preparations for the Granton railway, or the stratification of Salisbury Crags—always learning something new, and winning ideas for imagination to feed upon. One Saturday, February 12, 1842, he had a special treat, being taken 'to see electro-magnetic machines.'"

* "Life of J. C. Maxwell," p. 52.

And again, speaking of his school-life :—

"But at school also he gradually made his way. He soon discovered that Latin was worth learning, and the Greek Delectus interested him when we got so far. And there were two subjects in which he at once took the foremost place, when he had a fair chance of doing so; these were Scripture Biography and English. In arithmetic as well as in Latin his comparative want of readiness kept him down.

"On the whole he attained a measure of success which helped to secure for him a certain respect; and, however strange he sometimes seemed to his companions, he had three qualities which they could not fail to understand—agile strength of limb, imperturbable courage, and profound good-nature. Professor James Muirhead remembers him as 'a friendly boy, though never quite amalgamating with the rest.' And another old class-fellow, the Rev. W. Macfarlane of Lenzie, records the following as his impression: 'Clerk Maxwell, when he entered the Academy, was somewhat rustic and somewhat eccentric. Boys called him "Dafty," and used to try to make fun of him. On one occasion I remember he turned with tremendous vigour, with a kind of demonic force, on his tormentors. I think he was let alone after that, and gradually won the respect even of the most thoughtless of his schoolfellows.'"

The first reference to mathematical studies occurs, says Professor Campbell, in a letter to his father written soon after his thirteenth birthday.*

"After describing the Virginian Minstrels, and betwixt inquiries after various pets at Glenlair, he remarks, as if it were an ordinary piece of news, 'I have made a tetrahedron, a dodecahedron, and two other hedrons, whose names I don't know.' We had not yet begun geometry, and he had certainly not at this time learnt the definitions in Euclid; yet he had

---

\* "Life of J. C. Maxwell," p 56.

not merely realised the nature of the five regular solids sufficiently to construct them out of pasteboard with approximate accuracy, but had further contrived other symmetrical polyhedra derived from them, specimens of which (as improved in 1848) may be still seen at the Cavendish Laboratory.

"Who first called his attention to the pyramid, cube, etc., I do not know. He may have seen an account of them by chance in a book. But the fact remains that at this early time his fancy, like that of the old Greek geometers, was arrested by these types of complete symmetry; and his imagination so thoroughly mastered them that he proceeded to make them with his own hand. That he himself attached more importance to this moment than the letter indicates is proved by the care with which he has preserved these perishable things, so that they (or those which replaced them in 1848) are still in existence after thirty-seven years."

The summer holidays were spent at Glenlair. His cousin, Miss Jemima Wedderburn, was with him, and shared his play. Her skilled pencil has left us many amusing pictures of the time, some of which are reproduced by Professor Campbell. There were expeditions and picnics of all sorts, and a new toy known as "the devil on two sticks" afforded infinite amusement. The winter holidays usually found him at Penicuik, or occasionally at Glasgow, with Professor Blackburne or Professor W. Thomson (now Lord Kelvin). In October, 1844, Maxwell was promoted to the rector's class-room. John Williams, afterwards Archdeacon of Cardigan, a distinguished Baliol man, was rector, and the change was in many ways an important one for Maxwell. He writes to his father: "I like P—— better than B——. We have lots of jokes, and he speaks a great deal, and we have not so much monotonous parsing. In the English Milton is better than the History of Greece. . . ."

P—— was the boys' nickname for the rector; B—— for Mr. Carmichael, the second master. This * is the account of Maxwell's first interview with the rector :—

*Rector:* "What part of Galloway do you come from ?"

*J. C. M.:* "From the Vale of Urr. Ye spell it o, err, err, or oo, err, err."

The study of geometry was begun, and in the mathematical master, Mr. Gloag, Maxwell found a teacher with a real gift for his task. It was here that Maxwell's vast superiority to many who were his companions at once showed itself. "He seemed," says Professor Campbell, "to be in the heart of the subject when they were only at the boundary; but the boyish game of contesting point by point with such a mind was a most wholesome stimulus, so that the mere exercise of faculty was a pure joy. With Maxwell the first lessons of geometry branched out at once into inquiries which became fruitful."

In July, 1845, he writes :—

"I have got the 11th prize for Scholarship, the 1st for English, the prize for English verses, and the Mathematical Medal. I tried for Scripture knowledge, and Hamilton in the 7th has got it. We tried for the Medal on Thursday. I had done them all, and got home at half-past two ; but Campbell stayed till four. I was rather tired with writing exercises from nine till half-past two.

"Campbell and I went 'once more unto the b(r)each

* "Life of J. C. Maxwell," p. 67.

to-day at Portobello. I can swim a little now. Campbell has got 6 prizes. He got a letter written too soon, congratulating him upon *my* medal; but there is no rivalry betwixt us, as B——Carmichael says."

After a summer spent chiefly at Glenlair, he returned with his father to Edinburgh for the winter, and began, at the age of fourteen, to go to the meetings of the Royal Society of Edinburgh. At the Society of Arts he met Mr. R. D. Hay, the decorative painter, who had interested himself in the attempt to reduce beauty in form and colour to mathematical principles. Clerk Maxwell was interested in the question how to draw a perfect oval, and devised a method of drawing oval curves which was referred by his father to Professor Forbes for his criticism and suggestions. After discussing the matter with Professor Kelland, Professor Forbes wrote as follows *:—

"MY DEAR SIR,—I am glad to find to-day, from Professor Kelland, that his opinion of your son's paper agrees with mine, namely, that it is most ingenious, most creditable to him, and, we believe, a new way of considering higher curves with reference to foci. Unfortunately, these ovals appear to be curves of a very high and intractable order, so that possibly the elegant method of description may not lead to a corresponding simplicity in investigating their properties. But that is not the present point. If you wish it, I think that the simplicity and elegance of the method would entitle it to be brought before the Royal Society.—Believe me, my dear sir, yours truly, "JAMES D. FORBES.'

In consequence of this, Clerk Maxwell's first

* "Life of J. C. Maxwell," p. 75.

published paper was communicated to the Royal Society of Edinburgh on April 6th, 1846, when its author was barely fifteen. Its title is as follows: "On the Description of Oval Curves and those having a Plurality of Foci. By Mr. Clerk Maxwell, Junior. With Remarks by Professor Forbes. Communicated by Professor Forbes."

The notice in his father's diary runs: "M. 6 [Ap., 1846.] Royal Society with Jas. Professor Forbes gave acct. of James's Ovals. Met with very great attention and approbation generally."

This was the beginning of the lifelong friendship between Maxwell and Forbes.

The curves investigated by Maxwell have the property that the sum found by adding to the distance of any point on the curve from one focus a constant multiple of the distance of the same point from a second focus is always constant.

The curves are of great importance in the theory of light, for if this constant factor expresses the refractive index of any medium, then light diverging from one focus without the medium and refracted at a surface bounding the medium, and having the form of one of Maxwell's ovals, will be refracted so as to converge to the second focus.

About the same time he was busy with some investigations on the properties of jelly and guttapercha, which seem to have been suggested by Forbes' "Theory of Glaciers."

He failed to obtain the Mathematical Medal in 1846—possibly on account of these researches—but he continued at school till 1847, when he left, being then first in mathematics and in English, and nearly first in Latin.

In 1847 he was working at magnetism and the polarisation of light. Some time in that year he was taken by his uncle, Mr. John Cay, to see William Nicol, the inventor of the polarising prism, who showed him the colours exhibited by polarised light after passing through unannealed glass. On his return, he made a polariscope with a glass reflector. The framework of the first instrument was of cardboard, but a superior article was afterwards constructed of wood. Small lenses mounted on cardboard were employed when a conical pencil was needed. By means of this instrument he examined the figures exhibited by pieces of unannealed glass, which he prepared himself; and, with a camera lucida and box of colours, he reproduced these figures on paper, taking care to sketch no outlines, but to shade each coloured band imperceptibly into the next. Some of these coloured drawings he forwarded to Nicol, and was more than repaid by the receipt shortly afterwards of a pair of prisms prepared by Nicol himself. These prisms were always very highly prized by Maxwell. Once, when at Trinity, the little box containing them was carried off by his bed-maker during a vacation, and destined for destruction. The bed-maker died before term commenced, and it was only by diligent search among her effects that the prisms were recovered.* After this they were more carefully guarded, and they are now, together with the wooden polariscope, the bits of unannealed glass,

* Professor Garnett in *Nature*, November 13th, 1879.

and the water-colour drawings, in one of the showcases at the Cavendish Laboratory.

About this time, Professor P. G. Tait and he were schoolfellows at the Academy, acknowledged as the two best mathematicians in the school. It was thought desirable, says Professor Campbell, that " we should have lessons in physical science, so one of the classical masters gave them out of a text-book. . . . The only thing I distinctly remember about these hours is that Maxwell and P. G. Tait seemed to know much more about the subject than our teacher did."

An interesting account of these days is given by Professor Tait in an obituary notice on Maxwell printed in the "Proceedings of the Royal Society of Edinburgh, 1879-80," from which the following is taken:—

"When I first made Clerk Maxwell's acquaintance, about thirty-five years ago, at the Edinburgh Academy, he was a year before me, being in the fifth class, while I was in the fourth.

"At school he was at first regarded as shy and rather dull. He made no friendships, and he spent his occasional holidays in reading old ballads, drawing curious diagrams, and making rude mechanical models. This absorption in such pursuits, totally unintelligible to his schoolfellows (who were then quite innocent of mathematics), of course procured him a not very complimentary nickname, which I know is still remembered by many Fellows of this Society. About the middle of his school career, however, he surprised his companions by suddenly becoming one of the most brilliant among them, gaining high, and sometimes the highest, prizes for scholarships, mathematics, and English verse composition. From this time forward I became very intimate with him, and we discussed together, with schoolboy enthusiasm, numerous curious problems, among which I remember particularly the various plane sections of a ring or tore, and the form of a cylindrical mirror which should show one his own image unperverted. I still possess some of the MSS. we exchanged in 1846 and early in 1847. Those by Maxwell are on 'The Conical Pendulum,' 'Descartes' Ovals,' 'Meloid and Apioid,' and 'Trifocal Curves.' All are drawn up in strict geometrical form and divided into consecutive propositions. The three latter are connected with his first published paper, communicated by Forbes to this society and printed in our 'Proceedings,' vol. ii., under the title, 'On the Description of Oval Curves and those having a Plurality of Foci' (1846). At the time when these papers were written he had received no instruction in mathematics beyond a few books of Euclid and the merest elements of algebra."

In November, 1847, Clerk Maxwell entered the University of Edinburgh, learning mathematics from Kelland, natural philosophy from J. D. Forbes, and logic from Sir W. R. Hamilton. At this time, according to Professor Campbell *—

"he still occasioned some concern to the more conventional amongst his friends by the originality and simplicity of his ways. His replies in ordinary conversation were indirect and enigmatical, often uttered with hesitation and in a monotonous key. While extremely neat in his person, he had a rooted objection to the vanities of starch and gloves. He had a pious horror of destroying anything, even a scrap of writing-paper. He preferred travelling by the third class in railway journeys, saying he liked a hard seat. When at table he often seemed abstracted from what was going on, being absorbed in observing the effects of refracted light in the finger-glasses, or in trying some experiment with his eyes—seeing round a corner, making invisible stereoscopes, and the like. Miss Cay used to call his attention by crying, 'Jamsie, you're in a prop.' He never tasted wine; and he spoke to

* "Life of J. C. Maxwell," p. 105.

gentle and simple in exactly the same tone. On the other hand, his teachers—Forbes above all—had formed the highest opinion of his intellectual originality and force ; and a few experienced observers, in watching his devotion to his father, began to have some inkling of his heroic singleness of heart. To his college companions, whom he could now select at will, his quaint humour was an endless delight. His chief associates, after I went to the University of Glasgow, were my brother, Robert Campbell (still at the Academy), P. G. Tait, and Allan Stewart. Tait went to Peterhouse, Cambridge, in 1848, after one session of the University of Edinburgh ; Stewart to the same college in 1849 ; Maxwell did not go up until 1850."

During this period he wrote two important papers. The one, on "Rolling Curves," was read to the Royal Society of Edinburgh by Professor Kelland —(" it was not thought proper for a boy in a round jacket to mount the rostrum ")—in February, 1849 ; the other, on "The Equilibrium of Elastic Solids, appeared in the spring of 1850.

The vacations were spent at Glenlair, and we learn from letters to Professor Campbell and others how the time was passed.

"On Saturday," he writes*—April 26th, 1848, just after his arrival home—"the natural philosophers ran up Arthur's Seat with the barometer. The Professor set it down at the top. . . . He did not set it straight, and made the hill grow fifty feet ; but we got it down again."

In a letter of July in the same year he describes his laboratory :—

"I have regularly set up shop now above the wash-house at the gate, in a garret. I have an old door set on two barrels,

* " Life of J. C. Maxwell," p. 116.

and two chairs, of which one is safe, and a skylight above which will slide up and down.

"On the door (or table) there is a lot of bowls, jugs, plates, jam pigs, etc., containing water, salt, soda, sulphuric acid, blue vitriol, plumbago ore; also broken glass, iron, and copper wire, copper and zinc plate, bees' wax, sealing wax, clay, rosin, charcoal, a lens, a Smee's galvanic apparatus, and a countless variety of little beetles, spiders, and wood lice, which fall into the different liquids and poison themselves. I intend to get up some more galvanism in jam pigs; but I must first copper the interiors of the pigs, so I am experimenting on the best methods of electrotyping. So I am making copper seals with the device of a beetle. First, I thought a beetle was a good conductor, so I embedded one in wax (not at all cruel, because I slew him in boiling water, in which he never kicked), leaving his back out; but he would not do. Then I took a cast of him in sealing wax, and pressed wax into the hollow, and blackleaded it with a brush; but neither would that do. So at last I took my fingers and rubbed it, which I find the best way to use the blacklead. Then it coppered famously. I melt out the wax with the lens, that being the cleanest way of getting a strong heat, so I do most things with it that need heat. To-day I astonished the natives as follows. I took a crystal of blue vitriol and put the lens to it, and so drove off the water, leaving a white powder. Then I did the same to some washing soda, and mixed the two white powders together, and made a small native spit on them, which turned them green by a mutual exchange, thus:—1. Sulphate of copper and carbonate of soda. 2. Sulphate of soda and carbonate of copper (blue or green)."

Of his reading he says:—" I am reading Herodotus' 'Euterpe,' having taken the turn—that is to say that sometimes I can do props., read Diff. and Int. Calc., Poisson, Hamilton's dissertation, etc."

In September he was busy with polarised light. "We were at Castle Douglas yesterday, and got crystals of saltpetre, which I have been cutting up into plates to-day in hopes to see rings."

In July, 1849, he writes * : —

"I have set up the machine for showing the rings in crystals, which I planned during your visit last year. It answers very well. I also made some experiments on compressed jellies in illustration of my props. on that subject. The principal one was this:—The jelly is poured while hot into the annular space contained between a paper cylinder and a cork; then, when cold, the cork is twisted round and the jelly exposed to polarised light, when a transverse cross, X, not +, appears, with rings as the inverse square of the radius, all which is fully verified. Hip! etc. *Q.E.D.*"

And again on March 22nd, 1850:—

"At Practical Mechanics I have been turning Devils of sorts. For private studies I have been reading Young's 'Lectures,' Willis's 'Principles of Mechanism,' Moseley's 'Engineering and Mechanics,' Dixon on 'Heat,' and Moigno's 'Répertoire d'Optique.' This last is a very complete analysis of all that has been done in the optical way from Fresnel to the end of 1849, and there is another volume a coming which will complete the work. There is in it, besides common optics, all about the other things which accompany light, as heat, chemical action, photographic rays, action on vegetables, etc.

"My notions are rather few, as I do not *entertain* them just now. I have a notion for the torsion of wires and rods, not to be made till the vacation ; of experiments on the action of compression on glass, jelly, etc., numerically done up; of papers for the Physico-Mathematical Society (which is to revive in earnest next session!); on the relations of optical and mechanical constants, their desirableness, etc. ; and suspension bridges, and catenaries, and elastic curves. Alex. Campbell, Agnew, and I are appointed to read up the subject of periodical shooting stars, and to prepare a list of the phenomena to be observed on the 9th August and 13th

* "Life of J. C. Maxwell," pp. 123-129.

November. The society's barometer is to be taken up Arthur's Seat at the end of the session, when Forbes goes up, and All students are invited to attend, so that the existence of the society may be recognised."

It was at last settled that he was to go up to Cambridge. Tait had been at Peterhouse for two years, while Allan Stewart had joined him there in 1849, and after much discussion it was arranged that Maxwell should enter at the same college.

Of this period of his life Tait writes as follows :—

"The winter of 1847 found us together in the classes of Forbes and Kelland, where he highly distinguished himself. With the former he was a particular favourite, being admitted to the free use of the class apparatus for original experiments. He lingered here behind most of his former associates, having spent three years at the University of Edinburgh, working (without any assistance or supervision) with physical and chemical apparatus, and devouring all sorts of scientific works in the library. During this period he wrote two valuable papers, which are published in our 'Transactions,' on 'The Theory of Rolling Curves' and on 'The Equilibrium of Elastic Solids.' Thus he brought to Cambridge, in the autumn of 1850, a mass of knowledge which was really immense for so young a man, but in a state of disorder appalling to his methodical private tutor. Though that tutor was William Hopkins, the pupil to a great extent took his own way, and it may safely be said that no high wrangler of recent years ever entered the Senate House more imperfectly trained to produce 'paying' work than did Clerk Maxwell. But by sheer strength of intellect, though with the very minimum of knowledge how to use it to advantage under the conditions of the examination, he obtained the position of Second Wrangler, and was bracketed equal with the Senior Wrangler in the higher ordeal of the Smith's Prizes. His name appears in the Cambridge 'Calendar' as Maxwell of Trinity, but he was originally entered at Peterhouse, and kept his first term there, in that small but most ancient foundation which has of late furnished Scotland with the majority of the professors of mathematics and natural philosophy in her four universities."

While W. D. Niven, in his preface to Maxwell's collected works (p. xii.), says :—

"It may readily be supposed that his preparatory training for the Cambridge course was far removed from the ordinary type. There had indeed for some time been practically no restraint upon his plan of study, and his mind had been allowed to follow its natural bent towards science, though not to an extent so absorbing as to withdraw him from other pursuits. Though he was not a sportsman—indeed, sport so-called was always repugnant to him—he was yet exceedingly ond of a country life. He was a good horseman and a good swimmer. Whence, however, he derived his chief enjoyment may be gathered from the account which Mr. Campbell gives of the zest with which he quoted on one occasion the lines of Burns which describe the poet finding inspiration while wandering along the banks of a stream in the free indulgence of his fancies. Maxwell was not only a lover of poetry, but himself a poet, as the fine pieces gathered together by Mr. Campbell abundantly testify. He saw, however, that his true calling was science, and never regarded these poetical efforts as other than mere pastime. Devotion to science, already stimulated by successful endeavour; a tendency to ponder over philosophical problems; and an attachment to English literature, particularly to English poetry—these tastes, implanted in a mind of singular strength and purity, may be said to have been the endowments with which young Maxwell began his Cambridge career. Besides this, his scientific reading, as we may gather from his papers to the Royal Society of Edinburgh referred to above, was already extensive and varied. He brought with him, says Professor Tait, a mass of knowledge which was really immense for so young a man, but in a state of disorder appalling to his methodical private tutor."

## CHAPTER II.

### UNDERGRADUATE LIFE AT CAMBRIDGE.

MAXWELL did not remain long at Peterhouse; before the end of his first term he migrated to Trinity, and was entered under Dr. Thompson December 14th, 1850. He appeared to the tutor a shy and diffident youth, but presently surprised Dr. Thompson by producing a bundle of papers—copies, probably, of those he had already published — and remarking, "Perhaps these may show that I am not unfit to enter at your College."

The change was pressed upon him by many friends, the grounds of the advice being that, from the large number of high wranglers recently at Peterhouse and the smallness of the foundation, the chances of a Fellowship there for a mathematical man were less than at Trinity. It was a step he never regretted; the prospect of a Fellowship had but little influence on his mind. He found, however, at the larger college ampler opportunities for self-improvement, and it was possible for him to select his friends from among men whom he otherwise would never have known.

The record of his undergraduate life is not very full; his letters to his father have, unfortunately, been lost, but we have enough in the recollections of friends still living to picture what it was like. At first he lodged in King's Parade with an old Edinburgh schoolfellow, C. H. Robertson. He attended the

College lectures on mathematics, though they were somewhat elementary, and worked as a private pupil with Porter, of Peterhouse. His father writes to him, November, 1850: "Have you called on Professors Sedgwick, at Trin., and Stokes, at Pembroke? If not, you should do both. Stokes will be most in your line, if he takes you in hand at all. Sedgwick is also a great Don in his line, and, if you were entered in geology, would be a most valuable acquaintance."

In his second year he became a pupil of Hopkins, the great coach: he also attended Stokes' lectures, and the friendship which lasted till his death was thus begun. In April, 1852, he was elected a scholar, and obtained rooms in College (G. Old Court). In June, 1852, he came of age. "I trust you will be as discreet when major as you have been while minor," writes his father the day before. The next academic year, October, 1852, to June, 1853, was a very busy one; hard grind for the Tripos occupied his time, and he seems to have been thoroughly overstrained. He was taken ill while staying near Lowestoft with the Rev. C. B. Tayler, the uncle of a College friend. His own account of the illness is given in a letter to Professor Campbell *, dated July 14th, 1853.

"You wrote just in time for your letter to reach me as I reached Cambridge. After examination, I went to visit the Rev. C. B. Tayler (uncle to a Tayler whom I think you have seen under the name of *Freshman*, etc., and author of many tracts and other didactic works. We had little expedites and walks, and things parochial and educational, and domesticity. I intended to return on the 18th June, but on the 17th I felt

* "Life of J. C. Maxwell," p. 190.

unwell, and took measures accordingly to be well again—*i.e.*
went to bed, and made up my mind to recover. But it lasted
more than a fortnight, during which time I was taken care of
beyond expectation (not that I did not expect much before).
When I was perfectly useless and could not sit up without
fainting, Mr. Tayler did everything for me in such a way that
I had no fear of giving trouble. So did Mrs. Tayler; and the
two nephews did all they could. So they kept me in great
happiness all the time, and detained me till I was able to walk
about and got back strength. I returned on the 4th July.

"The consequence of all this is that I correspond with Mr.
Tayler, and have entered into bonds with the nephews, of
all of whom more hereafter. Since I came here I have been
attending Hop., but, with his approval, did not begin full
swing. I am getting on, though, and the work is not grinding
on the prepared brain."

During this period he wrote some papers for the
*Cambridge and Dublin Mathematical Journal* which
will be referred to again later. He was also a member
of a discussion society known as the "Apostles," and
some of the essays contributed by him are preserved
by Professor Campbell. Mr. Niven, in his preface to
the collected edition of Maxwell's works, suggests
that the composition of these essays laid the foundation of that literary finish which is one of the
characteristics of Maxwell's scientific writings.

Among his friends at the time were Tait, Charles
Mackenzie of Caius, the missionary bishop of Central
Africa, Henry and Frank Mackenzie of Trinity,
Droop, third Wrangler in 1854; Gedge, Isaac Taylor,
Blakiston, F. W. Farrar,[*] H. M. Butler,[†] Hort, V.
Lushington, Cecil Munro, G. W. H. Tayler, and W. N.
Lawson. Some of these who survived him have

---

[*] Dean of Canterbury. [†] Master of Trinity.

given to Professor Campbell their recollections of these undergraduate days, which are full of interest.

Thus Mr. Lawson writes * :—

"There must be many of his quaint verses about, if one could lay hands on them, for Maxwell was constantly producing something of the sort and bringing it round to his friends, with a sly chuckle at the humour, which, though his own, no one enjoyed more than himself.

"I remember Maxwell coming to me one morning with a copy of verses beginning, 'Gin a body meet a body going through the air,' in which he had twisted the well-known song into a description of the laws of impact of solid bodies.

"There was also a description which Maxwell wrote of some University ceremony—I forget what—in which somebody 'went before' and somebody 'followed after,' and 'in the midst were the wranglers, playing with the symbols.'

"These last words, however meant, were, in fact, a description of his own wonderful power. I remember, one day in lecture, our lecturer had filled the black-board three times with the investigation of some hard problem in Geometry of Three Dimensions, and was not at the end of it, when Maxwell came up with a question whether it would not come out geometrically, and showed how, with a figure, and in a few lines, there was the solution at once.

"Maxwell was, I daresay you remember, very fond of a talk upon almost anything. He and I were pupils at an enormous distance apart) of Hopkins, and I well recollect how, when I had been working the night before and all the morning at Hopkins's problems, with little or no result, Maxwell would come in for a gossip, and talk on while I was wishing him far away, till at last, about half an hour or so before our meeting at Hopkins's, he would say, 'Well, I must go to old Hop.'s problems'; and, by the time we met there, they were all done.

"I remember Hopkins telling me, when speaking of Maxwell, either just before or just after his degree, 'It is not

---

* " Life of J. C. Maxwell," p. 174.

possible for that man to think incorrectly on physical subjects'; and Hopkins, as you know, had had, perhaps, more experience of mathematical minds than any man of his time."

The last clause is part of a quotation from a diary kept by Mr. Lawson at Cambridge, in which, under the date July 15th, 1853, he writes :—

"He (Hopkins) was talking to me this evening about Maxwell. He says he is unquestionably the most extraordinary man he has met with in the whole range of his experience ; he says it appears impossible for Maxwell to think incorrectly on physical subjects; that in his analysis, however, he is far more deficient. He looks upon him as a great genius with all its eccentricities, and prophesies that one day he will shine as a light in physical science—a prophecy in which all his fellow-students strenuously unite."

How many who have struggled through the "Electricity and Magnetism" have realised the truth of the remark about the correctness of his physical intuitions and the deficiency at times of his analysis!

Dr. Butler, a friend of these early days, preached the University sermon on November 16th, 1879, ten days after Maxwell's death, and spoke thus :—

"It is a solemn thing—even the least thoughtful is touched by it—when a great intellect passes away into the silence and we see it no more. Such a loss, such a void, is present, I feel certain, to many here to-day. It is not often, even in this great home of thought and knowledge, that so bright a light is extinguished as that which is now mourned by many illustrious mourners, here chiefly, but also far beyond this place. I shall be believed when I say in all simplicity that I wish it had fallen to some more competent tongue to put into words those feelings of reverent affection which are, I am persuaded, uppermost in many hearts on this Sunday. My poor words shall be few, but believe me they come from the heart. You know, brethren, with what an eager pride we follow the fortunes of those whom we have loved and reverenced in our undergraduate days. We may see them but seldom, few letters may pass between us, but their names are never common names. They never become to us only what other men are. When I came up to Trinity twenty-eight years ago, James Clerk Maxwell was just beginning his second year. His position among us—I speak in the presence of many who remember that time—was unique. He was the one acknowledged man of genius among the undergraduates. We understood even then that, though barely of age, he was in his own line of inquiry not a beginner but a master. His name was already a familiar name to men of science. If he lived, it was certain that he was one of that small but sacred band to whom it would be given to enlarge the bounds of human knowledge. It was a position which might have turned the head of a smaller man; but the friend of whom we were all so proud, and who seemed, as it were, to link us thus early with the great outside world of the pioneers of knowledge, had one of those rich and lavish natures which no prosperity can impoverish, and which make faith in goodness easy for others. I have often thought that those who never knew the grand old Adam Sedgwick and the then young and ever-youthful Clerk Maxwell had yet to learn the largeness and fulness of the moulds in which some choice natures are framed. Of the scientific greatness of our friend we were most of us unable to judge; but anyone could see and admire the boy-like glee, the joyous invention, the wide reading, the eager thirst for truth, the subtle thought, the perfect temper, the unfailing reverence, the singular absence of any taint of the breath of worldliness in any of its thousand forms.

" Brethren, you may know such men now among your college friends, though there can be but few in any year, or indeed in any century, that possess the rare genius of the man whom we deplore. If it be so, then, if you will accept the counsel of a stranger, thank God for His gift. Believe me when I tell you that few such blessings will come to you in later life. There

c

are blessings that come once in a lifetime. One of these is the reverence with which we look up to greatness and goodness in a college friend—above us, beyond us, far out of our mental or moral grasp, but still one of us, near to us, our own. You know, in part at least, how in this case the promise of youth was more than fulfilled, and how the man who, but a fortnight ago, was the ornament of the University, and—shall I be wrong in saying it ?—almost the discoverer of a new world of knowledge, was even more loved than he was admired, retaining after twenty years of fame that mirth, that simplicity, that child-like delight in all that is fresh and wonderful which we rejoice to think of as some of the surest accompaniment of true scientific genius.

"You know, also, that he was a devout as well as thoughtful Christian. I do not note this in the triumphant spirit of a controversialist. I will not for a moment assume that there is any natural opposition between scientific genius and simple Christian faith. I will not compare him with others who have had the genius without the faith. Christianity, though she thankfully welcomes and deeply prizes them, does not need now, any more than when St. Paul first preached the Cross at Corinth, the speculations of the subtle or the wisdom of the wise. If I wished to show men, especially young men, the living force of the Gospel, I would take them not so much to a learned and devout Christian man to whom all stores of knowledge were familiar, but to some country village where for fifty years there had been devout traditions and devout practice. There they would see the Gospel lived out; truths, which other men spoke of, seen and known; a spirit not of this world, visibly, hourly present; citizenship in heaven daily assumed and daily realised. Such characters I believe to be the most convincing preachers to those who ask whether Revelation is a fable and God an unknowable. Yes, in most cases—not, I admit, in all simple faith, even peradventure more than devout genius, is mighty for removing doubts and implanting fresh conviction. But having said this, we may well give thanks to God that our friend was what he was, a firm Christian believer, and that his powerful mind, after ranging at will through the illimitable spaces of Creation and almost handling what he called 'the foundation-stones of the material universe,' found its true rest and happiness in the love and the mercy of Him whom the humblest Christian calls his Father. Of such a man it may be truly said that he had his citizenship in heaven, and that he looked for, as a Saviour, the Lord Jesus Christ, through whom the unnumbered worlds were made, and in the likeness of whose image our new and spiritual body will be fashioned."

The Tripos came in January, 1854. "You will need to get muffetees for the Senate Room. Take your plaid or rug to wrap round your feet and legs," was his father's advice—advice which will appeal to many who can remember the Senate House as it felt on a cold January morning.

Maxwell had been preparing carefully for this examination. Thus to his aunt, Miss Cay, in June, 1853, he writes:—"If anyone asks how I am getting on in mathematics, say that I am busy arranging everything so as to be able to express all distinctly, so that examiner may be satisfied now and pupils edified hereafter. It is pleasant work and very strengthening, but not nearly finished."

Still, the illness of July, 1853, had left some effect. Professor Baynes states that he said that on entering the Senate House for the first paper he felt his mind almost a blank, but by-and-by his mental vision became preternaturally clear.

The moderators were Mackenzie of Caius, whose advice had been mainly instrumental in leading him to migrate to Trinity, Wm. Walton of Trinity, Wolstenholme of Christ's, and Percival Frost of St. John's.

When the lists were published, Routh of Peterhouse was senior, Maxwell second. The examination for the Smith's Prizes followed in a few days, and then Routh and Maxwell were declared equal.

In a letter to Miss Cay * of January 13th, while waiting for the three days' list, he writes :—

"All my correspondents have been writing to me, which is kind, and have not been writing questions, which is kinder. So I answer you now, while I am slacking speed to get up steam, leaving Lewis and Stewart, etc., till next week, when I will give an account of the *five days*. There are a good many up here at present, and we get on very jolly on the whole; but some are not well, and some are going to be plucked or gulphed, as the case may be, and others are reading so hard that they are invisible. I go to-morrow to breakfast with shaky men, and after food I am to go and hear the list read out, and whether they are through, and bring them word. When the honour list comes out the poll men act as messengers. Bob Campbell comes in occasionally of an evening now, to discuss matters and vary sports. During examination I have had men at night working with gutta-percha, magnets, etc. It is much better than reading novels or talking after 5½ hours' hard writing."

His father, on hearing the news, wrote from Edinburgh :—

"I heartily congratulate you on your place in the list. I suppose it is higher than the speculators would have guessed, and quite as high as Hopkins reckoned on. I wish you success in the Smith's Prizes ; be sure to write me the result. I will see Mrs. Morrieson, and I think I will call on Dr. Gloag to congratulate him. He has at least three pupils gaining honours."

His friends in Edinburgh were greatly pleased,

* "Life of J. C. Maxwell," p. 195.

"I get congratulations on all hands," his father writes,* "including Professor Kelland and Sandy Fraser and all others competent. . . . To-night or on Monday I shall expect to hear of the Smith's Prizes." And again, February 6th, 1854:—"George Wedderburn came into my room at 2 a.m. yesterday morning, having seen the Saturday *Times*, received by the express train. . . . As you are equal to the Senior in the champion trial, you are very little behind him."

Or again, March 5th, 1854:—

"Aunt Jane stirred me up to sit for my picture, as she said you wished for it and were entitled to ask for it *quâ* Wrangler. I have had four sittings to Sir John Watson Gordon, and it is now far advanced; I think it is very like. It is kitcat size, to be a companion to Dyce's picture of your mother and self, which Aunt Jane says she is to leave to you.

And now the long years of preparation were nearly over. The cunning craftsman was fitted with his tools; he could set to work to unlock the secrets of Nature; he was free to employ his genius and his knowledge on those tasks for which he felt most fitted.

* "Life of J. C. Maxwell," p. 207.

## CHAPTER III.

### EARLY RESEARCHES.—PROFESSOR AT ABERDEEN.

FROM this time on Maxwell's life becomes a record of his writings and discoveries. It will, however, probably be clearest to separate as far as possible biographical details from a detailed account of his scientific work, leaving this for consecutive treatment in later chapters, and only alluding to it so far as may prove necessary to explain references in his letters.

He continued in Cambridge till the Long Vacation of 1854, reading Mill's "Logic." " I am experiencing the effects of Mill," he writes, March 25th, 1854, " but I take him slowly. I do not think him the last of his kind. I think more is wanted to bring the connexion of sensation with science to light, and to show what it is not." He also read Berkeley on "The Theory of Vision" and "greatly admired it."

About the same time he devised an ophthalmo-scope.*

" I have made an instrument for seeing into the eye through the pupil. The difficulty is to throw the light in at that small hole and look in at the same time; but that difficulty is overcome, and I can see a large part of the back of the eye quite distinctly with the image of the candle on it. People find no inconvenience in being examined, and I have got dogs to sit quite still and keep their eyes steady. Dogs' eyes are very beautiful behind—a copper-coloured ground, with

---

* " Life of J. C. Maxwell," p. 208.

glorious bright patches and networks of blue, yellow, and green, with blood-vessels great and small."

After the vacation he returned to Cambridge, and the letters refer to the colour-top. Thus to Miss Cay, November 24th, 1854, p. 208:—

"I have been very busy of late with various things, and am just beginning to make papers for the examination at Cheltenham, which I have to conduct about the 11th of December. I have also to make papers to polish off my pups with. I have been spinning colours a great deal, and have got most accurate results, proving that ordinary people's eyes are all made alike, though some are better than others, and that other people see two colours instead of three; but all those who do so agree amongst themselves. I have made a triangle of colours by which you may make out everything.

"If you can find out any people in Edinburgh who do not see colours (I know the Dicksons don't), pray drop a hint that I would like to see them. I have put one here up to a dodge by which he distinguishes colours without fail. I have also constructed a pair of squinting spectacles, and am beginning operations on a squinting man."

A paper written for his own use originally some time in 1854, but communicated as a parting gift to his friend Farrar, who was about to become a master at Marlborough, gives us some insight into his view of life at the age of twenty-three.

"He that would enjoy life and act with freedom must have the work of the day continually before his eyes. Not yesterday's work, lest he fall into despair; nor to-morrow's, lest he become a visionary—not that which ends with the day, which is a worldly work; nor yet that only which remains to eternity, for by it he cannot shape his actions.

"Happy is the man who can recognise in the work of to-day a connected portion of the work of life and an embodiment of the work of Eternity. The foundations of his confidence are unchangeable, for he has been made a partaker of Infinity. He strenuously works out his daily enterprises because the present is given him for a possession.

"Thus ought Man to be an impersonation of the divine process of nature, and to show forth the union of the infinite with the finite, not slighting his temporal existence, remembering that in it only is individual action possible; nor yet shutting out from his view that which is eternal, knowing that Time is a mystery which man cannot endure to contemplate until eternal Truth enlighten it."

His father was unwell in the Christmas vacation of that year, and he could not return to Cambridge at the beginning of the Lent term. "My steps," he writes* to C. J. Munro from Edinburgh, February 19th, 1855, "will be no more by the reedy and crooked till Easter term. . . . I should like to know how many kept bacalaurean weeks go to each of these terms, and when they begin and end. Overhaul the Calendar, and when found make note of."

He was back in Cambridge for the May term, working at the motion of fluids and at his colour-top. A paper on "Experiments on Colour as Perceived by the Eye" was communicated to the Royal Society of Edinburgh on March 19th, 1855. The experiments were shown to the Cambridge Philosophical Society in May following, and the results are thus described in two letters† to his father, Saturday, May 5th, 1855:

"The Royal Society have been very considerate in sending me my paper on 'Colours' just when I wanted it for the Philosophical here. I am to let them see the tricks on Monday

---

\* "Life of J. C. Maxwell," p. 210.
† "Life of J. C. Maxwell," p. 211.

evening, and I have been there preparing their experiments in the gaslight. There is to be a meeting in my rooms to-night to discuss Adam Smith's 'Theory of Moral Sentiments,' so I must clear up my litter presently. I am working away at electricity again, and have been working my way into the views of heavy German writers. It takes a long time to reduce to order all the notions one gets from these men, but I hope to see my way through the subject and arrive at something intelligible in the way of a theory. . . . .

"The colour trick came off on Monday, 7th. I had the proof-sheets of my paper, and was going to read; but I changed my mind and talked instead, which was more to the purpose. There were sundry men who thought that blue and yellow make green, so I had to undeceive them. I have got Hay's book of colours out of the Univ. Library, and am working through the specimens, matching them with the top. I have a new trick of stretching the string horizontally above the top, so as to touch the upper part of the axis. The motion of the axis sets the string a-vibrating in the same time with the revolutions of the top, and the colours are seen in the haze produced by the vibration. Thomson has been spinning the top, and he finds my diagram of colours agrees with his experiments, but he doubts about browns, what is their composition. I have got colcothar brown, and can make white with it, and blue and green : also, by mixing red with a little blue and green and a great deal of black, I can match colcothar exactly.

"I have been perfecting my instrument for looking into the eye. Ware has a little beast like old Ask, which sits quite steady and seems to like being looked at, and I have got several men who have large pupils and do not wish to let me look in. I have seen the image of the candle distinctly in all the eyes I have tried, and the veins of the retina were visible in some ; but the dogs' eyes showed all the ramifications of veins, with glorious blue and green network, so that you might copy down everything. I have shown lots of men the image in my own eye by shutting off the light till the pupil dilated and then letting it on.

"I am reading Electricity and working at Fluid Motion, and have got out the condition of a fluid being able to flow the same way for a length of time and not wriggle about."

The British Association met at Glasgow in September, 1855, and Maxwell was present, and showed his colour-top at Professor Ramsay's house to some of those interested. Letters* to his father about this time describe some of the events of the meeting and his own plans for the term.

"We had a paper from Brewster on 'The theory of three colours in the spectrum,' in which he treated Whewell with philosophic pity, commending him to the care of Prof. Wartman of Geneva, who was considered the greatest authority in cases of his kind—cases, in fact, of colour-blindness. Whewell was in the room, but went out and avoided the quarrel; and Stokes made a few remarks, stating the case not only clearly but courteously. However, Brewster did not seem to see that Stokes admitted his experiments to be correct, and the newspapers represented Stokes as calling in question the accuracy of the experiments.

"I am getting my electrical mathematics into shape, and I see through some parts which were rather hazy before; but I do not find very much time for it at present, because I am reading about heat and fluids, so as not to tell lies in my lectures. I got a note from the Society of Arts about the platometer, awarding thanks and offering to defray the expenses to the extent of £10, on the machine being produced in working order. When I have arranged it in my head, I intend to write to James Bryson about it.

"I got a long letter from Thomson about colours and electricity. He is beginning to believe in my theory about all colours being capable of reference to three standard ones, and he is very glad that I should poach on his electrical preserves.

". . . It is difficult to keep up one's interest in intel-

* "Life of J. C. Maxwell," p. 216.

lectual matters when friends of the intellectual kind are scarce. However, there are plenty friends not intellectual who serve to bring out the active and practical habits of mind, which overly-intellectual people seldom do. Wherefore, if I am to be up this term, I intend to addict myself rather to the working men who are getting up classes than to pups., who are in the main a vexation. Meanwhile, there is the examination to consider.

"You say Dr. Wilson has sent his book. I will write and thank him. I suppose it is about colour-blindness. I intend to begin Poisson's papers on electricity and magnetism tomorrow. I have got them out of the library. My reading hitherto has been of novels—'Shirley' and 'The Newcomes,' and now 'Westward Ho.'

"Macmillan proposes to get up a book of optics with my assistance, and I feel inclined for the job. There is great bother in making a mathematical book, especially on a subject with which you are familiar, for in correcting it you do as you would to pups.—look if the principle and result is right, and forget to look out for small errors in the course of the work. However, I expect the work will be salutary, as involving hard work, and in the end much abuse from coaches and students, and certainly no vain fame, except in Macmillan's puffs. But, if I have rightly conceived the plan of an educational book on optics, it will be very different in manner, though not in matter, from those now used."

The examination referred to was that for a Fellowship at Trinity, and Maxwell was elected on October 10th, 1855.

He was immediately asked to lecture for the College, on hydrostatics and optics, to the upper division of the third year, and to set papers for the questionists. In consequence, he declined to take pupils, in order to have time for reading and doing private mathematics, and for seeing the men who attended his lectures.

In November he writes: "I have been lecturing two weeks now, and the class seems improving; and they come and ask questions, which is a good sign. I have been making curves to show the relations of pressure and volume in gases, and they make the subject easier."

Still, he found time to attend Professor Willis's lectures on mechanism and to continue his reading. "I have been reading," he writes, "old books on optics, and find many things in them far better than what is new. The foreign mathematicians are discovering for themselves methods which were well known at Cambridge in 1720, but are now forgotten."

The "Poisson" was read to help him with his own views on electricity, which were rapidly maturing, and the first of that great series of works which has revolutionised the science was published on December 10th, 1855, when his paper on "Faraday's Lines of Force" was read to the Cambridge Philosophical Society.

The next term found him back in Cambridge at work on his lectures, full of plans for a new colour top and other matters. Early in February he received a letter from Professor Forbes, telling him that the Professorship of Natural Philosophy in Marischal College, Aberdeen, was vacant, and suggesting that he should apply.

He decided to be a candidate if his father approved. "For my own part," he writes, "I think the sooner I get into regular work the better, and that the best way of getting into such work is to profess one's readiness by applying for it." On the

20th of February he writes: " However, wisdom is of many kinds, and I do not know which dwells with wise counsellors most, whether scientific, practical, political, or ecclesiastical. I hear there are candidates of all kinds relying on the predominance of one or other of these kinds of wisdom in the constitution of the Government."

The second part of the paper on "Faraday's Lines of Force" was read during the term. Writing on the 4th of March, he expresses the hope soon to be able to write out fully the paper. "I have done nothing in that way this term," he says, " but am just beginning to feel the electrical state come on again."

His father was working at Edinburgh in support of his candidature for Aberdeen, and when, in the middle of March, he returned North, he found everything well prepared. The two returned to Glenlair together after a few days in Edinburgh, and Maxwell was preparing to go back to Cambridge, when, on the 2nd of April, his father died suddenly.

Writing to Mrs. Blackburn, he says: " My father died suddenly to-day at twelve o'clock. He had been giving directions about the garden, and he said he would sit down and rest a little, as usual. After a few minutes I asked him to lie down on the sofa, and he did not seem inclined to do so; and then I got him some ether, which had helped him before. Before he could take any he had a slight struggle, and all was over. He hardly breathed afterwards."

Almost immediately after this, Maxwell was appointed to Aberdeen. His father's death had frustrated some at least of the intentions with which he had applied for the post. He knew the old man would be glad to see him the occupant of a Scotch chair. He hoped, too, to be able to live with his father at Glenlair for one half the year; but this was not to be. No doubt the laboratory and the freedom of the post, when compared with the routine work of preparing men for the Tripos, had their inducements; still, it may be doubted if the choice was a wise one for him. The work of drilling classes, composed, for the most part, of raw untrained lads, in the elements of physics and mechanics was, as Niven says in his preface to the collected works, not that for which he was best fitted; while at Cambridge, had he stayed, he must always have had among his pupils some of the best mathematicians of the time; and he might have founded some ten or fifteen years before he did that Cambridge School of Physicists which looks back with so much pride to him as their master.

Leave-taking at Trinity was a sad task. He writes * thus, June 4th, to Mr. R. B. Litchfield :—

"On Thursday evening I take the North-Western route to the North. I am busy looking over immense rubbish of papers, etc., for some things not to be burnt lie among much combustible matter, and some is soft and good for packing.

"It is not pleasant to go down to live solitary, but it would not be pleasant to stay up either, when all one had to do lay elsewhere. The transition state from a man into a Don must come at last, and it must be painful, like gradual outrooting of nerves. When it is done there is no more pain, but occasional reminders from some suckers, tap-roots, or other remnants of the old nerves, just to show what was there and what might have been."

\* "Life of J. C. Maxwell," p. 256.

The summer of 1856 was spent at Glenlair, where various friends were his guests—Lushington, MacLennan, the two cousins Cay, and others. He continued to work at optics, electricity, and magnetism, and in October was busy with "a solemn address or manifesto to the Natural Philosophers of the North, which needed coffee and anchovies and a roaring hot fire and spread coat-tails to make it natural." This was his inaugural lecture.

In November he was at Aberdeen. Letters[*] to Miss Cay, Professor Campbell, and C. J. Munro tell of the work of the session. The last is from Glenlair, dated May 20th, 1857, after work was over.

"The session went off smoothly enough. I had Sun, all the beginning of optics, and worked off all the experimental part up to Fraunhofer's lines, which were glorious to see with a water-prism I have set up in the form of a cubical box, five inch side. . . .

"I succeeded very well with heat. The experiments on latent heat came out very accurate. That was my part, and the class could explain and work out the results better than I expected. Next year I intend to mix experimental physics with mechanics, devoting Tuesday and THURSDAY (what would Stokes say?) to the science of experimenting accurately. . . .

"Last week I brewed chlorophyll (as the chemists word it), a green liquor, which turns the invisible light red. . . .

"My last grind was the reduction of equations of colour which I made last year. The result was eminently satisfactory."

Another letter,[†] June 5th, 1857, also to Munro, refers to the work of the University Commission and the new statutes.

[*] " Life of J. C. Maxwell," p. 267.
[†] " Life of J. C. Maxwell," p. 269.

"I have not seen Article 7, but I agree with your dissent from it entirely. On the vested interest principle, I think the men who intended to keep their fellowships by celibacy and ordination, and got them on that footing, should not be allowed to desert the virgin choir or neglect the priestly office, but on those principles should be allowed to live out their days, provided the whole amount of souls cured annually does not amount to £20 in the King's Book. But my doctrine is that the various grades of College officers should be set on such a basis that, although chance lecturers might be sometimes chosen from among fresh fellows who are going away soon, the reliable assistant tutors, and those that have a plain calling that way, should, after a few years, be elected permanent officers of the College, and be tutors and deans in their time, and seniors also, with leave to marry, or, rather, never prohibited or asked any questions on that head, and with leave to retire after so many years' service as seniors. As for the men of the world, we should have a limited term of existence, and that independent of marriage or 'parsonage.'"

It was more than twenty years before the scheme outlined in the above letter came to anything; but, at the time of Maxwell's death in 1879, another Commission was sitting, and the plan suggested by Maxwell became the basis of the statutes of nearly all the colleges.

For the winter session of 1857-58 he was again at Aberdeen.

The Adams Prize had been established in 1848 by some members of St. John's College, and connected by them with the name of Adams "in testimony of their sense of the honour he had conferred upon his College and the University by having been the first among the mathematicians of Europe to determine from perturbations the unknown place of a disturbing planet exterior to Uranus." Professor Challis, Dr. Parkinson, and Sir William Thomson, the examiners, had selected as the subject for the prize to be awarded in 1857 the "Motions of Saturn's Rings." For this Maxwell had decided to compete, and his letters at the end of 1857 tell of the progress of the task. Thus, writing * to Lewis Campbell from Glenlair on August 28th, he says:—

"I have been battering away at Saturn, returning to the charge every now and then. I have effected several breaches in the solid ring, and now I am splash into the fluid one, amid a clash of symbols truly astounding. When I reappear it will be in the dusky ring, which is something like the state of the air supposing the siege of Sebastopol conducted from a forest of guns 100 miles one way, and 30,000 miles the other, and the shot never to stop, but go spinning away round a circle, radius 170,000 miles."

And again † to Miss Cay on the 28th of November:—

"I have been pretty steady at work since I came. The class is small and not bright, but I am going to give them plenty to do from the first, and I find it a good plan. I have a large attendance of my old pupils, who go on with the higher subjects. This is not part of the College course, so they come merely from choice, and I have begun with the least amusing part of what I intend to give them. Many had been reading in summer, for they did very good papers for me on the old subjects at the beginning of the month. Most of my spare time I have been doing Saturn's rings, which is getting on now, but lately I have had a great many long letters to write —some to Glenlair, some to private friends, and some all about science. . . . I have had letters from Thomson and Challis about Saturn—from Hayward, of Durham University, about

* " Life of J. C. Maxwell," p. 278.
† " Life of J. C. Maxwell," p. 292.

the brass top, of which he wants one. He says that the earth has been really found to change its axis regularly in the way I supposed. Faraday has also been writing about his own subjects. I have had also to write Forbes a long report on colours; so that for every note I have got I have had to write a couple of sheets in reply, and reporting progress takes a deal of writing and spelling.

He devised a model (now at the Cavendish Laboratory) to exhibit the motions of the satellites in a disturbed ring, "for the edification of sensible image-worshippers."

The essay was awarded the prize, and secured for its author great credit among scientific men.

In another letter, written during the same session, he says: "I find my principal work here is teaching my men to avoid vague expressions, as 'a certain force,' meaning uncertain; *may* instead of *must*; *will be* instead of *is*; *proportional* instead of *equal*."

The death, during the autumn, of his College friend Pomeroy, from fever in India, was a great blow to him; his letters at the time show the depth of his feelings and his beliefs.

The question of the fusion of the two Colleges at Aberdeen, King's College and the Marischal College, was coming to the fore. "Know all men," he says, in a letter to Professor Campbell, "that I am a Fusionist."

In February, 1858, he was still engaged on Saturn's rings, while hard at work during the same time with his classes. He had established a voluntary class for his students of the previous year, and was reading with them Newton's "Lunar Theory and Astronomy." This was followed by "Electricity and Magnetism,"

Faraday's book being the backbone of everything, "as he himself is the nucleus of everything electric since 1830."

In February, 1858, he announced his engagement to Katherine Mary Dewar, the daughter of the Principal of Marischal College.

"Dear Aunt" (he says,* February 18th, 1858), "this comes to tell you that I am going to have a wife. . . .

"Don't be afraid ; she is not mathematical, but there are other things besides that, and she certainly won't stop mathematics. The only one that can speak as an eye-witness is Johnnie, and he only saw her when we were both trying to act the indifferent. We have been trying it since, but it would not do, and it was not good for either."

The wedding took place early in June. Professor Campbell has preserved some of the letters written by Maxwell to Miss Dewar, and these contain "the record of feelings which in the years that followed were transfused in action and embodied in a married life which can only be spoken of as one of unexampled devotion."

The project for the fusion of the two Colleges, to which reference has been made, went on, and the scheme was completed in 1860.

The two Colleges were united to form the University of Aberdeen, and the new chair of Natural Philosophy thus created was filled by the appointment of David Thomson, Professor of Natural Philosophy in King's College, and Maxwell's senior. Mr. W. D. Niven, in his preface to Maxwell's works, when dealing with this appointment, writes :—

* "Life of J. C. Maxwell," p. 303.

"Professor Thomson, though not comparable to Maxwell as a physicist, was nevertheless a remarkable man. He was distinguished by singular force of character and great administrative faculty, and he had been prominent in bringing about the fusion of the Colleges. He was also an admirable lecturer and teacher, and had done much to raise the standard of scientific education in the north of Scotland. Thus the choice made by the Commissioners, though almost inevitable, had the effect of making it appear that Maxwell failed as a teacher. There seems, however, to be no evidence to support such an inference. On the contrary, if we may judge from the number of voluntary students attending his classes in his last College session, he would seem to have been as popular as a professor as he was personally estimable."

The question whether Maxwell was a great teacher has sometimes been discussed. I trust that the following pages will give an answer to it. He was not a prominent lecturer. As Professor Campbell says,* "Between his students' ignorance and his vast knowledge it was difficult to find a common measure. The advice which he once gave to a friend whose duty it was to preach to a country congregation, 'Why don't you give it them thinner?' must often have been applicable to himself. . . . Illustrations of *ignotum per ignotius*, or of the abstruse by some unobserved property of the familiar, were multiplied with dazzling rapidity. Then the spirit of indirectness and paradox, though he was aware of its dangers, would often take possession of him against his will, and, either from shyness or momentary excitement, or the despair of making himself understood, would land him in 'chaotic

* "Life of J. C. Maxwell," p. 259.

statements,' breaking off with some quirk of ironical humour."

But teaching is not all done by lecturing. His books and papers are vast storehouses of suggestions and ideas which the ablest minds of the past twenty years have been since developing. To talk with him for an hour was to gain inspiration for a year's work; to see his enthusiasm and to win his praise or commendation were enough to compensate for many weary struggles over some stubborn piece of apparatus which would not go right, or some small source of error which threatened to prove intractable and declined to submit itself to calculation. The sure judgment of posterity will confirm the verdict that Clerk Maxwell was a great teacher, though lecturing to a crowd of untrained undergraduates was a task for which others were better fitted than he.

## CHAPTER IV.

### PROFESSOR AT KING'S COLLEGE, LONDON.—LIFE AT GLENLAIR.

In 1860 Forbes resigned the chair of Natural Philosophy at Edinburgh. Maxwell and Tait were candidates, and Tait was appointed. In the summer of the same year Maxwell obtained the vacant Professorship of Natural Philosophy at King's College, London. This he held to 1865, and this period of his life is distinguished by the appearance of some of his most important papers. The work was arduous; the College course extended over nine months of the year; there were as well evening lectures to artisans as part of his regular duties. His life in London was useful to him in the opportunities it gave him for becoming personally acquainted with Faraday and others. He also renewed his intimacy with various Cambridge friends.

He was at the celebrated Oxford meeting of the British Association in 1860, where he exhibited his colour-box for mixing the colours of the spectrum. In 1859, at the meeting at Aberdeen, he had read to Section A his first paper on the "Dynamical Theory of Gases," published in the *Philosophical Magazine* for January, 1860. The second part of the paper, dealing with the conduction of heat and other phenomena in a gas, was published in July, 1860, after the Oxford meeting.

A paper on the "Theory of Compound Colours"

was communicated to the Royal Society by Professor Stokes in January, 1860. It contains the account of his colour-box in the form finally adopted (most of the important parts of the apparatus are still at the Cavendish Laboratory), and a number of observations by Mrs. Maxwell and himself, which will be more fully described later.

In November, 1860, he received for this work the Rumford medal of the Royal Society.

The next year, 1861, is of great importance in the history of electrical science. The British Association met at Manchester, and a Committee was appointed on Standards of Electrical Resistance. Maxwell was not a member. The committee reported at the Cambridge meeting in 1862, and were reappointed with extended duties. Maxwell's name, among others, was added, and he took a prominent part in the deliberations of the committee, which, as their Report* presented in 1863 states, came to the opinion, "after mature consideration, that the system of so-called absolute electrical units, based on purely mechanical measurements, is not only the best system yet proposed, but is the only one consistent with our present knowledge both of the relations existing between the various electrical phenomena and of the connection between these and the fundamental measurements of time, space, and mass."

Appendix C of this Report, "On the Elementary Relations between Electrical Measurements," bears the names of Clerk Maxwell and Fleeming Jenkin, and is the foundation of everything that has been done in

* B.A. Report, Newcastle, 1863.

the way of absolute electrical measurement since that date; while Appendix D gives an account by the same two workers of the experiments on the absolute unit of electrical resistance made in the laboratory of King's College by Maxwell, Fleeming Jenkin, and Balfour Stewart. Further experiments are described in the report for 1864. The work thus begun was consummated during the year 1894 by the legalisation throughout the civilised world of a system of electrical units based on those described in these reports.

Meanwhile, Maxwell's views on electro-magnetic theory were quietly developing. Papers on "Physical Lines of Force," which appeared in the *Philosophical Magazine* during 1861 and 1862, contain the germs of his theory—expressed at that time, it is true, in a somewhat material form. In the paper published January, 1862, the now well-known relation between the ratio of the electric units and the velocity of light was established, and his correspondence with Fleeming Jenkin and C. J. Munro about this time relates in part to the experimental verification of this relation. His experiments on this matter were published in the "Philosophical Transactions" for 1868.

This electrical theory occupied his mind mainly during 1863 and 1864. In September of the latter year he writes [*] from Glenlair to C. Hockin, who had taken Balfour Stewart's place during the second series of experiments on the measurement of resistance.

"I have been doing several electrical problems. I have got a theory of 'electric absorption,' *i.e.*, residual charge, etc., and I very much want determinations of the specific induction,

[*] "Life of J. C. Maxwell," p. 340.

electric resistance, and absorption of good dielectrics, such as glass, shell-lac, gutta-percha, ebonite, sulphur, etc.

"I have also cleared the electromagnetic theory of light from all unwarrantable assumption, so that we may safely determine the velocity of light by measuring the attraction between bodies kept at a given difference of potential, the value of which is known in electromagnetic measure.

"I hope there will be resistance coils at the British Association."

This work resulted in his greatest electrical paper, "A Dynamical Theory of the Electromagnetic Field," read to the Royal Society December 8th, 1864.

But the molecular theory of gases was still prominently before his mind.

In 1862, writing * to H. R. Droop, he says :—

"Some time ago, when investigating Bernoulli's theory of gases, I was surprised to find that the internal friction of a gas (if it depends on the collision of particles) should be independent of the density.

"Stokes has been examining Graham's experiments on the rate of flow of gases through fine tubes, and he finds that the friction, if independent of density, accounts for Graham's results; but, if taken proportional to density, differs from those results very much. This seems rather a curious result, and an additional phenomenon, explained by the 'collision of particles' theory of gases. Still one phenomenon goes against that theory—the relation between specific heat at constant pressure and at constant volume, which is in air $= 1\cdot408$, while it ought to be $1\cdot333$."

And again† in the same year, 21st April, 1862, to Lewis Campbell :—

"Herr Clausius of Zürich, one of the heat philosophers, has been working at the theory of gases being little bodies flying

* "Life of J. C. Maxwell," p. 332.
† "Life of J. C. Maxwell," p. 336.

about, and has found some cases in which he and I don't tally. So I am working it out again. Several experimental results have turned up lately rather confirmatory than otherwise of that theory.

"I hope you enjoy the absence of pupils. I find the division of them into smaller classes is a great help to me and to them; but the total oblivion of them for definite intervals is a necessary condition for doing them justice at the proper time."

The experiments on the viscosity of gases, which formed the Bakerian Lecture to the Royal Society read on February 8th, 1866, were the outcome of this work. His house in 8, Palace Gardens, Kensington, contained a large garret running the complete length.

"To maintain the proper temperature a large fire was for some days kept up in the room in the midst of very hot weather. Kettles were kept on the fire and large quantities of steam allowed to flow into the room. Mrs. Maxwell acted as stoker, which was very exhausting work when maintained for several consecutive hours. After this the room was kept cool for subsequent experiments by the employment of a considerable amount of ice."

Next year, May, 1866, was read his paper on the "Dynamical Theory of Gases," in which errors in his former papers, which had been pointed out by Clausius, were corrected.

Meanwhile he had resigned his London Professorship at the end of the Session of 1865, and had been succeeded by Professor W. G. Adams.

For the next four years he lived chiefly at Glenlair, working at his theory of electricity, occasionally, as we shall see, visiting London and Cambridge, and taking an active interest in the affairs of his own neighbourhood. In 1865 he had a serious illness, through which he was nursed with great care by Mrs. Maxwell. His correspondence was considerable, and absorbed much of his time. Much also was given to the study of English literature; he was fond of reading Chaucer, Milton, or Shakespeare aloud to Mrs. Maxwell.

He also read much theological and philosophical literature, and all he read helped only to strengthen that firm faith in the fundamentals of Christianity in which he lived and died.

In 1867 he and Mrs. Maxwell paid a visit to Italy, which was a source of great pleasure to both.

His chief scientific work was the preparation of his "Electricity and Magnetism," which did not appear till 1873; the time was in the main one of quiet thought and preparation for his next great task, the foundation of the School of Physics in Cambridge.

In 1868 the principalship of the United College in the University of St. Andrews was vacant by the resignation of Forbes, and Maxwell was invited by several of the professors to stand. He, however, declined to submit his name to the Crown.

## CHAPTER V.

### CAMBRIDGE.—PROFESSOR OF PHYSICS.

DURING his retirement at Glenlair from 1865 to 1870 Maxwell was frequently at Cambridge. He examined in the Mathematical Tripos in 1866 and 1867, and again in 1869 and 1870.

The regulations for the Tripos had been in force practically unchanged since 1848, and it was felt by many that the range of subjects included was not sufficiently extensive, and that changes were urgently needed if Cambridge were to retain its position as the centre of mathematical teaching. Natural Philosophy was mentioned in the Schedule, but Natural Philosophy included only Dynamics and Astronomy, Hydrostatics and Physical Optics, with some simple Hydrodynamics and Sound.

The subjects of Heat, Electricity and Magnetism, the Theory of Elastic Solids and Vibrations, Vortex-Motion in Hydrodynamics, and much else, were practically new since 1848. Stokes, Thomson, and Maxwell in England, and Helmholtz in Germany, had created them.

Accordingly in June, 1868, a new plan of examinations was sanctioned by the Senate to come into force in January, 1873, and these various subjects were explicitly included.

Mr. Niven, who was one of those examined by Maxwell in 1866, writes in the preface to the collected works :—

"For some years previous to 1866, when Maxwell returned to Cambridge as Moderator in the Mathematical Tripos, the studies in the University had lost touch with the great scientific movements going on outside her walls. It was said that some of the subjects most in vogue had but little interest for the present generation, and loud complaints began to be heard that while such branches of knowledge as Heat, Electricity, and Magnetism were left out of the Tripos examination, the candidates were wasting their time and energy upon mathematical trifles barren of scientific interest and of practical results. Into the movement for reform Maxwell entered warmly. By his questions in 1866, and subsequent years, he infused new life into the examination; he took an active part in drafting the new scheme introduced in 1873; but most of all by his writings he exerted a powerful influence on the younger members of the University, and was largely instrumental in bringing about the change which has been now effected."

But the University possessed no means of teaching these subjects, and a Syndicate or Committee was appointed, November 25th, 1868, " to consider the best means of giving instruction to students in Physics, especially in Heat, Electricity and Magnetism, and the methods of providing apparatus for this purpose."

Dr. Cookson, Master of St. Peter's College, took an active part in the work of the Syndicate. Professor Stokes, Professor Liveing, Professor Humphry, Dr. Phear, and Dr. Routh were among the members. Maxwell himself was in Cambridge that winter, as Examiner for the Tripos, and his work as Moderator and Examiner in the two previous years had done much to show the necessity of alterations and to indicate the direction which changes should take.

The Syndicate reported February 27th, 1869. They called attention to the Report of the Royal Commission of 1850. The Commissioners had "prominently urged the importance of cultivating a knowledge of the great branches of Experimental Physics in the University"; and in page 118 of their Report, after commending the manner in which the subject of Physical Optics is studied in the University, and pointing out that "there is, perhaps, no public institution where it is better represented or prosecuted with more zeal and success in the way of original research," they had stated that "no reason can be assigned why other great branches of Natural Science should not become equally objects of attention, or why Cambridge should not become a great school of physical and experimental, as it is already of mathematical and classical, instruction."

And again the Commissioners remark: "In a University so thoroughly imbued with the mathematical spirit, physical study might be expected to assume within its precincts its highest and severest tone, be studied under more abstract forms, with more continual reference to mathematical laws, and therefore with better hope of bringing them one by one under the domain of mathematical investigation than elsewhere."

After calling attention to these statements the Report of the Syndicate then continues:—

"In the scheme of Examination for Honours in the Mathematical Tripos approved by Grace of the Senate on the 2nd of June, 1868, Heat, Electricity and Magnetism, if not introduced for the first time, had a much greater degree of importance assigned to them than at any previous period, and these subjects will henceforth demand a corresponding amount of attention from the candidates for Mathematical Honours. The Syndicate have limited their attention almost entirely to the question of providing public instruction in Heat, Electricity and Magnetism. They recognise the importance and advantage of tutorial instruction in these subjects in the several colleges, but they are also alive to the great impulse given to studies of this kind, and to the large amount of additional training which students may receive through the instruction of a public Professor, and by knowledge gained in a well-appointed laboratory."

"In accordance with these views, and at an early period in their deliberations, they requested the Professors* of the University, who are engaged in teaching Mathematical and Physical Science, to confer together upon the present means of teaching Experimental Physics, especially Heat, Electricity and Magnetism, and to inform them how the increased requirements of the University in this respect could be met by them."

"The Professors, so consulted, favoured the Syndicate with a report on the subject, which the Syndicate now beg leave to lay before the Senate. It points out how the requirements of the University might be "partially met," but the Professors state distinctly that they "do not think that they are able to meet the want of an extensive course of lectures on Physics

---

* The Professors who were consulted were Challis, Willis, Stokes, Cayley, **Adams**, and Liveing.

treated as such, and in great measure experimentally. As Experimental Physics may fairly be considered to come within the province of one or more of the above-mentioned Professors, the Syndicate have considered whether now or at some future time some arrangement might not be made to secure the effective teaching of this branch of science, without having resort to the services of an additional Professor. They are, however, of opinion that such an arrangement cannot be made at the present time, and that the exigencies of the case may be best met by founding a new professorship which shall terminate with the tenure of office of the Professor first elected. The services of a man of the highest attainments in science, devoting his life to public teaching as such Professor, and engaged in original research, would be of incalculable benefit to the University."

The Report goes on to point out that a laboratory would be necessary, and also apparatus. It is estimated that £5,000 would cover the cost of the laboratory, and £1,300 the necessary apparatus. Provision is also made for a demonstrator and a laboratory assistant, and the Report closes with a recommendation that a special Syndicate of Finance should be appointed to consider the means of raising the funds.

The Professors in their Report to the Syndicate point out that teaching in Experimental Physics is needed for the Mathematical Tripos, the Natural Sciences Tripos, certain Special examinations, and the first examination for the degree of M.B. It appeared to them clear that there was work for a new Professor.

In May, 1869, the Financial Syndicate recommended by the above Report was appointed "to consider the means of raising the necessary funds for establishing a professor and demonstrator of Experimental Physics, and for providing buildings and apparatus required for that department of science, and further to consider other wants of the University, and the sources from which those wants may be supplied."

The Syndicate endeavoured to meet the expenditure by inquiry from the several Colleges whether they would be willing to make contributions from their corporate funds, but without success.

"The answers of the Colleges indicated such a want of concurrence in any proposal to raise contributions from the corporate funds of Colleges by any kind of direct taxation that the Syndicate felt obliged to abandon the notion of obtaining the necessary funds from this source, and accordingly to limit the number of objects which they should recommend the Senate to accomplish."

External authority was necessary before the colleges would submit to taxation for University purposes, and it was left to the Royal Commission of 1877 to carry into effect many of the suggestions made by the Syndicate. Meanwhile they contented themselves with recommending means for raising an annual stipend of £660 for the professor, demonstrator, and assistant, and a capital sum of £5,000, or thereabouts, for the expenses of a building.

The Syndicate's Report was issued in an amended form in the May term of 1870, and before any decision was taken on it the Vice-Chancellor, Dr. Atkinson, on

October 13th, 1870, published "the following munificent offer of his grace the Duke of Devonshire, the Chancellor of the University," who had been chairman of the Commission on Scientific Education.

> "Holker Hall,
> Grange, Lancashire.
>
> "MY DEAR MR. VICE-CHANCELLOR,—I have the honour to address you for the purpose of making an offer to the University, which, if you see no objection, I shall be much obliged to you to submit in such manner as you may think fit for the consideration of the Council and the University.
>
> "I find in the report dated February 29th, 1869, of the Physical Science Syndicate, recommending the establishment of a Professor and Demonstrator of Experimental Physics, that the buildings and apparatus required for this department of science are estimated to cost £6,300.
>
> "I am desirous to assist the University in carrying this recommendation into effect, and shall accordingly be prepared to provide the funds required for the building and apparatus as soon as the University shall have in other respects completed its arrangements for teaching Experimental Physics, and shall have approved the plan of the building.
>
> "I remain, my dear Mr. Vice-Chancellor,
> "Yours very faithfully,
> "DEVONSHIRE."

By his generous action the University was relieved from all expense connected with the building. A Grace establishing a Professorship of Experimental Physics was confirmed by the Senate February 9th, 1871, and March 8th was fixed for the election.

Meanwhile who was to be Professor? Sir W. Thomson's name had been mentioned, but he, it was known, would not accept the post. Maxwell was then applied to, and at first he was unwilling to leave Glenlair. Professor Stokes, the Hon. J. W. Strutt

(Lord Rayleigh), Mr. Blore of Trinity, and others wrote to him. Lord Rayleigh's letter * is as follows:

"Cambridge, 14th February, 1871.
"When I came here last Friday I found everyone talking about the new professorship, and hoping that you would come. Thomson, it seems, has definitely declined. . . . There is no one here in the least fit for the post. What is wanted by most who know anything about it is not so much a lecturer as a mathematician who has actual experience in experimenting, and who might direct the energies of the younger Fellows and bachelors into a proper channel. There must be many who would be willing to work under a competent man, and who, while learning themselves, would materially assist him. . . . I hope you may be induced to come; if not, I don't know who it is to be. Do not trouble to answer me about this, as I believe others have written to you about it."

On the 15th of February, Maxwell wrote to Mr. Blore:—

"I had no intention of applying for the post when I got your letter, and I have none now, unless I come to see that I can do some good by it." The letter continues:—
"The class of Physical Investigations, which might be undertaken with the help of men of Cambridge education, and which would be creditable to the University, demand in general a considerable amount of dull labour, which may or may not be attractive to the pupils."

However, on the 24th of February, Mr. Blore wrote to the Electoral Roll:—

"I am authorised to give notice that Mr. John (*sic*) Clerk Maxwell, F.R.S., formerly Professor of Natural Philosophy at Aberdeen, and at King's College, London, is a candidate for the professorship of Experimental Physics."

* "Life of J. C. Maxwell," p. 349.

Maxwell was elected without opposition. Writing* to his wife from Cambridge, 20th March, 1871, he says :—

"There are two parties about the professorship. One wants popular lectures, and the other cares more for experimental work. I think there should be a gradation—popular lectures and rough experiments for the masses; real experiments for real students; and laborious experiments for first-rate men like Trotter and Stuart and Strutt."

While in a letter † from Glenlair to C. J. Munro, dated March 15th, 1871, he writes :—"The Experimental Physics at Cambridge is not built yet, but we are going to try. The desideratum is to set a Don and a Freshman to observe and register (say) the vibrations of a magnet together, or the Don to turn a watch and the Freshman to observe and govern him."

In October he delivered his Introductory Lecture. A few quotations will show the spirit in which he approached his task.

"In a course of Experimental Physics we may consider either the Physics or the Experiments as the leading feature. We may either employ the experiments to illustrate the phenomena of a particular branch of Physics, or we may make some physical research in order to exemplify a particular experimental method. In the order of time, we should begin, in the Lecture Room, with a course of lectures on some branch of Physics aided by experiments of illustration, and conclude, in the Laboratory, with a course of experiments of research.

"Let me say a few words on these two classes of experiments — Experiments of Illustration and Experiments of Research. The aim of an experiment of illustration is to

\* " Life of J. C. Maxwell," p. 381.
† " Life of J. C. Maxwell," p. 379.

throw light upon some scientific idea so that the student may be enabled to grasp it. The circumstances of the experiment are so arranged that the phenomenon which we wish to observe or to exhibit is brought into prominence, instead of being obscured and entangled among other phenomena, as it is when it occurs in the ordinary course of nature. To exhibit illustrative experiments, to encourage others to make them, and to cultivate in every way the ideas on which they throw light, forms an important part of our duty. The simpler the materials of an illustrative experiment, and the more familiar they are to the student, the more thoroughly is he likely to acquire the idea which it is meant to illustrate. The educational value of such experiments is often inversely proportional to the complexity of the apparatus. The student who uses home-made apparatus, which is always going wrong, often learns more than one who has the use of carefully adjusted instruments, to which he is apt to trust, and which he dares not take to pieces.

"It is very necessary that those who are trying to learn from books the facts of physical science should be enabled by the help of a few illustrative experiments to recognise these facts when they meet with them out of doors. Science appears to us with a very different aspect after we have found out that it is not in lecture-rooms only, and by means of the electric light projected on a screen, that we may witness physical phenomena, but that we may find illustrations of the highest doctrines of science in games and gymnastics, in travelling by land and by water, in storms of the air and of the sea, and wherever there is matter in motion.

"If, therefore, we desire, for our own advantage and for the honour of our University, that the Devonshire Laboratory should be successful, we must endeavour to maintain it in living union with the other organs and faculties of our learned body. We shall therefore first consider the relation in which we stand to those mathematical studies which have so long flourished among us, which deal with our own subjects, and which differ from our experimental studies only in the mode in which they are presented to the mind.

"There is no more powerful method for introducing knowledge into the mind than that of presenting it in as many different ways as we can. When the ideas, after entering through different gateways, effect a junction in the citadel of the mind, the position they occupy becomes impregnable. Opticians tell us that the mental combination of the views of an object which we obtain from stations no further apart than our two eyes is sufficient to produce in our minds an impression of the solidity of the object seen ; and we find that this impression is produced even when we are aware that we are really looking at two flat pictures placed in a stereoscope. It is therefore natural to expect that the knowledge of physical science obtained by the combined use of mathematical analysis and experimental research will be of a more solid, available, and enduring kind than that possessed by the mere mathematician or the mere experimenter.

"But what will be the effect on the University if men pursuing that course of reading which has produced so many distinguished Wranglers turn aside to work experiments ? Will not their attendance at the Laboratory count not merely as time withdrawn from their more legitimate studies, but as the introduction of a disturbing element, tainting their mathematical conceptions with material imagery, and sapping their faith in the formulæ of the text-books ? Besides this, we have already heard complaints of the undue extension of our studies, and of the strain put upon our questionists by the weight of learning which they try to carry with them into the Senate-House. If we now ask them to get up their subjects not only by books and writing, but at the same time by observation and manipulation, will they not break down altogether ? The Physical Laboratory, we are told, may perhaps be useful to those who are going out in Natural Science, and who do not take in Mathematics, but to attempt to combine both kinds of study during the time of residence at the University is more than one mind can bear.

"No doubt there is some reason for this feeling. Many of us have already overcome the initial difficulties of mathematical training. When we now go on with our study, we feel that it requires exertion and involves fatigue, but we are confident that if we only work hard our progress will be certain.

"Some of us, on the other hand, may have had some experience of the routine of experimental work. As soon as we can read scales, observe times, focus telescopes, and so on, this kind of work ceases to require any great mental effort. We may, perhaps, tire our eyes and weary our backs, but we do not greatly fatigue our minds.

"It is not till we attempt to bring the theoretical part of our training into contact with the practical that we begin to experience the full effect of what Faraday has called 'mental inertia'—not only the difficulty of recognising, among the concrete objects before us, the abstract relation which we have learned from books, but the distracting pain of wrenching the mind away from the symbols to the objects, and from the objects back to the symbols. This, however, is the price we have to pay for new ideas.

"But when we have overcome these difficulties, and successfully bridged over the gulph between the abstract and the concrete, it is not a mere piece of knowledge that we have obtained; we have acquired the rudiment of a permanent mental endowment. When, by a repetition of efforts of this kind, we have more fully developed the scientific faculty, the exercise of this faculty in detecting scientific principles in nature, and in directing practice by theory, is no longer irksome, but becomes an unfailing source of enjoyment, to which we return so often that at last even our careless thoughts begin to run in a scientific channel.

"Our principal work, however, in the Laboratory must be to acquaint ourselves with all kinds of scientific methods, to compare them and to estimate their value. It will, I think, be a result worthy of our University, and more likely to be accomplished here than in any private laboratory, if, by the free and full discussion of the relative value of different scientific procedures, we succeed in forming a school of scientific criticism and in assisting the development of the doctrine of method.

"But admitting that a practical acquaintance with the

methods of Physical Science is an essential part of a mathematical and scientific education, we may be asked whether we are not attributing too much importance to science altogether as part of a liberal education.

"Fortunately, there is no question here whether the University should continue to be a place of liberal education, or should devote itself to preparing young men for particular professions. Hence, though some of us may, I hope, see reason to make the pursuit of science the main business of our lives, it must be one of our most constant aims to maintain a living connexion between our work and the other liberal studies of Cambridge, whether literary, philological, historical, or philosophical.

"There is a narrow professional spirit which may grow up among men of science just as it does among men who practise any other special business. But surely a University is the very place where we should be able to overcome this tendency of men to become, as it were, granulated into small worlds, which are all the more worldly for their very smallness? We lose the advantage of having men of varied pursuits collected into one body if we do not endeavour to imbibe some of the spirit even of those whose special branch of learning is different from our own.'

Another expression of his views on the position of Physics at the time will be found in his address to Section A of the British Association, when President at the Liverpool meeting of 1870.

## CHAPTER VI.
### CAMBRIDGE—THE CAVENDISH LABORATORY.

BUT the laboratory was not yet built. A Syndicate, of which Maxwell was a member, was appointed to consider the question of a site, to take professional advice, and to obtain plans and estimates. Professor Maxwell and Mr. Trotter visited various laboratories at home and abroad for the purpose of ascertaining the best arrangements. Mr. W. M. Fawcett was appointed architect; the tender of Mr. John Loveday, of Kebworth, for the building at a cost of £8,450, exclusive of gas, water, and heating, was accepted in March, 1872, and the building* was begun during the summer.

In the meantime Maxwell began to lecture, finding a home where he could.

"Lectures begin 24th," he writes from Glenlair, October 19th, 1872. "Laboratory rising, I hear, but I have no place to erect my chair, but move about like the cuckoo, depositing my notions in the Chemical Lecture-room 1st term; in the Botanical in Lent, and in Comparative Anatomy in Easter."

It was not till June, 1874, that the building was complete, and on the 16th the Chancellor formally presented his gift of the Cavendish Laboratory to the University. In the correspondence previous to this time it was spoken of as the Devonshire Laboratory. The name Cavendish commemorated the work of the great physicist of a century earlier, whose writings

* An account of the laboratory is given in *Nature*, vol. x., p. 139.

Maxwell was shortly to edit, as well as the generosity of the Chancellor.

In their letter of thanks to the Duke of Devonshire the University write :—

"Unde vero conventius poterat illis artibus succurri quam e tua domo quæ in ipsis jam pridem inclaruerat. Notum est Henricum Cavendish quem secutus est Coulombius primum ita docuisse, quæ sit vis electrica ut eam numerorum modulis illustraret; adhibitis rationibus quas hodie veras esse constat." And they suggest the name as suitable for the building. To this the Chancellor replied, after referring to the work of Henry Cavendish: "Quod pono in officinâ ipsâ nuncupandâ nomen ejus commemorare dignati sitis, id grato animo accepi."

The building had cost far more than the original estimate, but the Chancellor's generosity was not limited, and on July 21st, 1874, he wrote to the Vice-Chancellor:—

"It is my wish to provide all instruments for the Cavendish Laboratory which Professor Maxwell may consider to be immediately required, either in his lectures or otherwise."

Maxwell prepared a list, but explained while doing it that time and thought were necessary to secure the best form of instruments; and he continues, writing to the Vice-Chancellor: "I think the Duke fully understood from what I said to him that to furnish the Laboratory will be a matter of several years' duration. I shall consider myself, however," he says, "at liberty to contribute to the Laboratory any instruments which I have had constructed in former years, and which may be found still useful, and also from time to time to procure others for special researches."

In 1877 in his annual report Professor Maxwell announced that the Chancellor\* had now "completed his gift to the University by furnishing the Cavendish Laboratory with apparatus suited to the present state of science."

The stock of apparatus, however, was still small, although Maxwell in the most generous manner himself spent large sums in adding to it; for the Professor was most particular in procuring only expensive instruments by the best makers, with such additional improvements as he could himself suggest.

In March, 1874, a Demonstratorship of Physics had been established, and Mr. Garnett of St. John's College was appointed.

Work began in the laboratory in October, 1874. At first the number of students was small. Only seventeen names appear in the Natural Sciences Tripos† list for 1874, and few of those did Physics.

The fear alluded to by the Professor in his introductory lecture, that men reading for the Mathematical

---

\* The Chancellor continued to take to the end of his life a warm interest in the work at the laboratory. In 1887, the Jubilee year, as Proctor—at the same time I held the office of Demonstrator—it was my duty to accompany the Chancellor and other officers to Windsor to present an address from the University to Her Majesty. I was introduced to the Chancellor at Paddington, and he at once began to question me closely about the progress of the laboratory, the number of students, and the work being done there, showing himself fully acquainted with recent progress.

† In 1894 the list contained, in Part II., sixteen names, and in Part I., one hundred and three names.

Tripos would not find time for attendance at the laboratory, was justified. One of the weaknesses of our Cambridge plan has been the divorce between Mathematics and experimental work, encouraged by our system of examinations. Experimental knowledge is supposed not to be needed for the Mathematical Tripos; the Mathematics permitted in the Natural Sciences Tripos are very simple; thus it came about that few men while reading for the Mathematical Tripos attended the laboratory, and this unfortunate result was intensified by the action of the University in 1877–78, when the regulations for the Mathematical Tripos were again altered.*

Still there were pupils eager and willing to work, though they were chiefly men who had already taken their B.A. degree, and who wished to continue Physical reading and research, even though it involved "a considerable amount of dull labour not altogether attractive." My own work there began in 1876, and it may be interesting if I recall my reminiscences of that time.

The first experiments I can recollect related to the measurement of electrical resistance. I well remember

* Under the new regulations Physics was removed from the first part of the Tripos and formed, with the more advanced parts of Astronomy and Pure Mathematics, a part by itself, to which only the Wranglers were admitted. Thus the number of men encouraged to read Physics was very limited. This pernicious system was altered in the regulations at present in force, which came into action in 1892. Part I. of the Mathematical Tripos now contains Heat, Elementary Hydrodynamics and Sound, and the simpler parts of Electricity and Magnetism, and candidates for this examination do come to the laboratory, though not in very large numbers. The more advanced parts both of Mathematics and Physics are included in Part II.

Maxwell explaining the principle of Wheatstone's bridge, and my own wish at the time that I had come to the laboratory before the Tripos, instead of afterwards. Lord Rayleigh had, during the examination, set an easy question which I failed to do for want of some slight experimental knowledge, and the first few words of Maxwell's talk showed me the solution.

I did not attend his lectures regularly—they were given, I think, at an hour which I was obliged to devote to teaching; besides, there was his book, the "Electricity and Magnetism," into which I had just dipped before the Tripos, to work at.

Chrystal and Saunder were then busy at their verification of Ohm's law. They were using a number of the Thomson form of tray Daniell's cells, and Maxwell was anxious for tests of various kinds to be made on these cells; these I undertook, and spent some time over various simple measurements on them. He then set me to work at some of the properties of a stratified dielectric, consisting, if I remember rightly, of sheets of paraffin paper and mica. By this means I became acquainted with various pieces of apparatus. There were no regular classes and no set drill of demonstrations arranged for examination purposes; these came later. In Maxwell's time those who wished to work had the use of the laboratory and assistance and help from him, but they were left pretty much to themselves to find out about the apparatus and the best methods of using it.

Rather later than this Schuster came and did some of his spectroscope work. J. E. H. Gordon was busy with the preliminary observations for his determination of Verdet's constant, and Niven had various electrical experiments on hand; while Fleming was at work on the B. A. resistance coils.

My own tastes lay in the direction of optics. Maxwell was anxious that I should investigate the properties of certain crystals. I think they were the chlorate of potash crystals, about which Stokes and Rayleigh have since written; but these crystals were to be grown, a slow process which would, he supposed, take years; and as I wished to produce a dissertation for the Trinity Fellowship examination in 1877, that work had to be laid aside.

Eventually I selected as a subject the form of the wave surface in a biaxial crystal, and set to work in a room assigned to me. The Professor used to come in on most days to see how I was getting on. Generally he brought his dog, which sometimes was shut up in the next room while he went to college. Dogs were not allowed in college, and Maxwell had an amusing way of describing how Toby once wandered into Trinity, and by some doggish instinct discovered immediately, to his intense amazement, that he was in a place where no dogs had been since the college was. Toby was not always quiet in his master's absence, and his presence in the next room was somewhat disturbing.

When difficulties occurred Maxwell was always ready to listen. Often the answer did not come at once, but it always did come after a little time. I remember one day, when I was in a serious dilemma, I told him my long tale, and he said:—

"Well, Chrystal has been talking to me, and

Garnett and Schuster have been asking questions, and all this has formed a good thick crust round my brain. What you have said will take some time to soak through, but we will see about it." In a few days he came back with—"I have been thinking over what you said the other day, and if you do so-and-so it will be all right."

My dissertation was referred to him, and on the day of the election, when returning to Cambridge for the admission, I met him at Bletchley station, and well remember his kind congratulations and words of warm encouragement.

For the next year and a half I was working regularly at the laboratory and saw him almost daily during term time.

Of these last years there really is but little to tell. His own scientific work went on. The "Electricity and Magnetism" was written mostly at Glenlair. About the time of his return to Cambridge, in October, 1872, he writes * to Lewis Campbell :—

"I am continually engaged in stirring up the Clarendon Press, but they have been tolerably regular for two months. I find nine sheets in thirteen weeks is their average. Tait gives me great help in detecting absurdities. I am getting converted to quaternions, and have put some in my book."

The book was published in 1873. The Text-book of Heat was written during the same period, while "Matter and Motion," "a small book on a great subject," was published in 1876.

In 1873 and 1874 he was one of the examiners for the Natural Sciences Tripos, and in 1873 he was the

* "Life of J. C. Maxwell," p. 383.

first additional examiner for the Mathematical Tripos, in accordance with the scheme which he had done so much to promote in 1868.

Many of his shorter papers were written about the same time. The ninth edition of the *Encyclopædia Britannica* was being published, and Professor Baynes had enlisted his aid in the work. The articles "Atom," "Attraction," "Capillary Action," "Constitution of Bodies," "Diffusion," "Ether," "Faraday," and others are by him.

He also wrote a number of papers for *Nature*. Some of these are reviews of books or accounts of scientific men, such as the notices of Faraday and Helmholtz, which appeared with their portraits; others again are original contributions to science. Among the latter many have reference to the molecular constitution of bodies. Two lectures—the first on "Molecules," delivered before the British Association at Bradford in 1873; the second on the "Dynamical Evidence of the Molecular Constitution of Bodies," delivered before the Chemical Society in 1875—were of special importance. The closing sentences of the first lecture have been often quoted. They run as follow:—

"In the heavens we discover by their light, and by their light alone, stars so distant from each other that no material thing can ever have passed from one to another; and yet this light, which is to us the sole evidence of the existence of these distant worlds, tells us also that each of them is built up of molecules of the same kinds as those which we find on earth. A molecule of hydrogen, for example, whether in Sirius or in Arcturus, executes its vibrations in precisely the same time.

"Each molecule therefore throughout the universe bears

impressed upon it the stamp of a metric system, as distinctly as does the metre of the Archives at Paris, or the double royal cubit of the temple of Karnac.

"No theory of evolution can be formed to account for the similarity of molecules, for evolution necessarily implies continuous change, and the molecule is incapable of growth or decay, of generation or destruction.

"None of the processes of Nature, since the time when Nature began, have produced the slightest difference in the properties of any molecule. We are therefore unable to ascribe either the existence of the molecules or the identity of their properties to any of the causes which we call natural.

"On the other hand, the exact equality of each molecule to all others of the same kind gives it, as Sir John Herschel has well said, the essential character of a manufactured article, and precludes the idea of its being eternal and self-existent.

"Thus we have been led along a strictly scientific path, very near to the point at which Science must stop—not that Science is debarred from studying the internal mechanism of a molecule which she cannot take to pieces any more than from investigating an organism which she cannot put together. But in tracing back the history of matter, Science is arrested when she assures herself, on the one hand, that the molecule has been made, and, on the other, that it has not been made by any of the processes we call natural.

"Science is incompetent to reason upon the creation of matter itself out of nothing. We have reached the utmost limits of our thinking faculties when we have admitted that because matter cannot be eternal and self-existent, it must have been created.

"It is only when we contemplate, not matter in itself, but the form in which it actually exists, that our mind finds something on which it can lay hold.

"That matter, as such, should have certain fundamental properties, that it should exist in space and be capable of motion, that its motion should be persistent, and so on, are truths which may, for anything we know, be of the kind which metaphysicians call necessary. We may use our knowledge of

F

such truths for purposes of deduction, but we have no data for speculating as to their origin.

"But that there should be exactly so much matter and no more in every molecule of hydrogen is a fact of a very different order. We have here a particular distribution of matter—a *collocation*, to use the expression of Dr. Chalmers, of things which we have no difficulty in imagining to have been arranged otherwise.

"The form and dimensions of the orbits of the planets, for instance, are not determined by any law of nature, but depend upon a particular collocation of matter. The same is the case with respect to the size of the earth, from which the standard of what is called the metrical system has been derived. But these astronomical and terrestrial magnitudes are far inferior in scientific importance to that most fundamental of all standards which forms the base of the molecular system. Natural causes, as we know, are at work which tend to modify, if they do not at length destroy, all the arrangements and dimensions of the earth and the whole solar system. But though in the course of ages catastrophes have occurred and may yet occur in the heavens, though ancient systems may be dissolved and new systems evolved out of their ruins, the molecules out of which these systems are built—the foundation stones of the material universe—remain unbroken and unworn. They continue this day as they were created—perfect in number and measure and weight; and from the ineffaceable characters impressed on them we may learn that those aspirations after accuracy in measurement, and justice in action, which we reckon among our noblest attributes as men, are ours because they are essential constituents of the image of Him who in the beginning created, not only the heaven and the earth, but the materials of which heaven and earth consist."

This was criticised in *Nature* by Mr. C. J. Munro, and at a later time by Clifford in one of his essays.

Some correspondence with the Bishop of Gloucester and Bristol on the authority for the comparison of molecules to manufactured articles is given by Professor Campbell, and in it Maxwell points out that the latter part of the article "Atom" in the *Encyclopædia* is intended to meet Mr. Munro's criticism.

In 1874 the British Association met at Belfast, under the presidency of Tyndall. Maxwell was present, and published afterwards in *Blackwood's Magazine* an amusing paraphrase of the president's address. This, with some other verses written at about the same time, may be quoted here. Professor Campbell has collected a number of verses written by Maxwell at various times, which illustrate in an admirable manner both the grave and the gay side of his character.

## BRITISH ASSOCIATION, 1874.

### *Notes of the President's Address.*

In the very beginnings of science, the parsons, who managed things then,
Being handy with hammer and chisel, made gods in the likeness of men;
Till commerce arose, and at length some men of exceptional power
Supplanted both demons and gods by the atoms, which last to this hour.
Yet they did not abolish the gods, but they sent them well out of the way,
With the rarest of nectar to drink, and blue fields of nothing to sway.
From nothing comes nothing, they told us— naught happens by chance, but by fate;
There is nothing but atoms and void, all else is mere whims out of date!
Then why should a man curry favour with beings who cannot exist,
To compass some petty promotion in nebulous kingdoms of mist?
But not by the rays of the sun, nor the glittering shafts of the day,
Must the fear of the gods be dispelled, but by words, and their wonderful play.

So treading a path all untrod, the poet-philosopher sings
Of the seeds of the mighty world—the first-beginnings of things;
How freely he scatters his atoms before the beginning of years;
How he clothes them with force as a garment, those small incompressible spheres!
Nor yet does he leave them hard-hearted—he dowers them with love and with hate,
Like spherical small British Asses in infinitesimal state;
Till just as that living Plato, whom foreigners nickname Plateau,*
Drops oil in his whisky-and-water (for foreigners sweeten it so);
Each drop keeps apart from the other, enclosed in a flexible skin,
Till touched by the gentle emotion evolved by the prick of a pin:
Thus in atoms a simple collision excites a sensational thrill,
Evolved through all sorts of emotion, as sense, understanding, and will
(For by laying their heads all together, the atoms, as councillors do,
May combine to express an opinion to every one of them new).
There is nobody here, I should say, has felt true indignation at all,
Till an indignation meeting is held in the Ulster Hall;
Then gathers the wave of emotion, then noble feelings arise,
Till you all pass a resolution which takes every man by surprise.
Thus the pure elementary atom, the unit of mass and of thought,
By force of mere juxtaposition to life and sensation is brought;
So, down through untold generations, transmission of structureless germs
Enables our race to inherit the thoughts of beasts, fishes, and worms.
We honour our fathers and mothers, grandfathers and grandmothers too;
But how shall we honour the vista of ancestors now in our view?
First, then, let us honour the atom, so lively, so wise, and so small;
The atomists next let us praise, Epicurus, Lucretius, and all.
Let us damn with faint praise Bishop Butler, in whom many atoms combined
To form that remarkable structure it pleased him to call—his mind.
Last, praise we the noble body to which, for the time, we belong,
Ere yet the swift whirl of the atoms has hurried us, ruthless, along,
The British Association—like Leviathan worshipped by Hobbes,
The incarnation of wisdom, built up of our witless nobs,
Which will carry on endless discussions when I, and probably you,
Have melted in infinite azure—in English, till all is blue.

---

* "Statique Expérimentale et Théorique des Liquides soumis aux seules Forces Moléculaires." Par J. Plateau, Professeur à l'Université de Gand.

## MOLECULAR EVOLUTION.
### Belfast, 1874.

At quite uncertain times and places,
 The atoms left their heavenly path,
And by fortuitous embraces
 Engendered all that being hath.
And though they seem to cling together,
 And form "associations" here,
Yet, soon or late, they burst their tether,
 And through the depths of space career.

So we who sat, oppressed with science,
 As British Asses, wise and grave,
Are now transformed to wild Red Lions,*
 As round our prey we ramp and rave.
Thus, by a swift metamorphosis,
 Wisdom turns wit, and science joke,
Nonsense is incense to our noses,
 For when Red Lions speak they smoke.

Hail, Nonsense! dry nurse of Red Lions,†
 From thee the wise their wisdom learn;
From thee they cull those truths of science,
 Which into thee again they turn.
What combinations of ideas
 Nonsense alone can wisely form!
What sage has half the power that she has,
 To take the towers of Truth by storm?

Yield, then, ye rules of rigid reason!
 Dissolve, thou too, too solid sense!
Melt into nonsense for a season,
 Then in some nobler form condense.
Soon, all too soon, the chilly morning
 This flow of soul will crystallise;
Then those who Nonsense now are scorning
 May learn, too late, where wisdom lies.

---

\* The "Red Lions" are a club formed by Members of the British Association to meet for relaxation after the graver labours of the day.

† "Lecnum arida nutrix." *Horace*.

## TO THE COMMITTEE OF THE CAYLEY PORTRAIT FUND.
### 1874.

O wretched race of men, to space confined!
What honour can ye pay to him, whose mind
    To that which lies beyond hath penetrated?
The symbols he hath formed shall sound his praise,
And lead him on through unimagined ways
    To conquests new, in worlds not yet created.

First, ye Determinants! in ordered row
And massive column ranged, before him go,
    To form a phalanx for his safe protection.
Ye powers of the $n^{th}$ roots of $-1$!
Around his head in ceaseless * cycles run,
    As unembodied spirits of direction.

And you, ye undevelopable scrolls!
Above the host wave your emblazoned rolls,
    Ruled for the record of his bright inventions.
Ye cubic surfaces! by threes and nines
Draw round his camp your seven-and-twenty lines —
    The seal of Solomon in three dimensions.

March on, symbolic host! with step sublime,
Up to the flaming bounds of Space and Time!
    There pause, until by Dickinson depicted,
In two dimensions, we the form may trace
Of him whose soul, too large for vulgar space,
    In $n$ dimensions flourished unrestricted.

---

## IN MEMORY OF EDWARD WILSON,

*Who repented of what was in his mind to write after section.*

    Rigid Body (*sings*).
    Gin a body meet a body
        Flyin' through the air,
    Gin a body hit a body,
        Will it fly? and where?

\* *v.r.*, endless.

> Ilka impact has its measure,
>   Ne'er a ane hae I ;
> Yet a' the lads they measure me,
>   Or, at least, they try.
>
> Gin a body meet a body
>   Altogether free,
> How they travel afterwards
>   We do not always see.
> Ilka problem has its method
>   By analytics high ;
> For me, I ken na ane o' them,
>   But what the waur am I ?

Another task, which occupied much time, from 1874 to 1879, was the edition of the works of Henry Cavendish. Cavendish, who was great-uncle to the Chancellor, had published only two electrical papers, but he had left some twenty packets of manuscript on Mathematical and Experimental Electricity. These were placed in Maxwell's hands in 1874 by the Duke of Devonshire.

Niven, in his preface to the collected papers dealing with this book, writes thus :—

"This work, published in 1879, has had the effect of increasing the reputation of Cavendish, disclosing as it does the unsuspected advances which that acute physicist had made in the Theory of Electricity, especially in the measurement of electrical quantities. The work is enriched by a variety of valuable notes, in which Cavendish's views and results are examined by the light of modern theory and methods. Especially valuable are the methods applied to the determination of the electrical capacities of conductors and condensers, a subject in which Cavendish himself showed considerable skill both of a mathematical and experimental character.

"The importance of the task undertaken by Maxwell in connection with Cavendish's papers will be understood from the following extract from his introduction to them :—

"'It is somewhat difficult to account for the fact that though Cavendish had prepared a complete description of his experiments on the charges of bodies, and had even taken the trouble to write out a fair copy, and though all this seems to have been done before 1774, and he continued to make experiments in electricity till 1781, and lived on till 1810, he kept his manuscript by him and never published it.

"'Cavendish cared more for investigation than for publication. He would undertake the most laborious researches in order to clear up a difficulty which no one but himself could appreciate or was even aware of, and we cannot doubt that the result of his enquiries, when successful, gave him a certain degree of satisfaction. But it did not excite in him that desire to communicate the discovery to others, which in the case of ordinary men of science generally ensures the publication of their results. How completely these researches of Cavendish remained unknown to other men of science is shown by the external history of electricity.'

"It will probably be thought a matter of some difficulty to place oneself in the position of a physicist of a century ago, and to ascertain the exact bearing of his experiments. But Maxwell entered upon this undertaking with the utmost enthusiasm, and succeeded in identifying himself with Cavendish's methods. He showed that Cavendish had really anticipated several of the discoveries in electrical science which have been made since his time. Cavendish was the first to form the conception of and to measure Electrostatic Capacity and Specific Inductive Capacity; he also anticipated Ohm's law."

During the last years of his life Mrs. Maxwell had a serious and prolonged illness, and Maxwell's work was much increased by his duties as sick nurse. On one occasion he did not sleep in a bed for three weeks, but conducted his lectures and experiments at the laboratory as usual.

About this time some of those who had been "Apostles" in 1853-57 revived the habit of meeting together for discussion. The club, which included Professors Lightfoot, Hort and Westcott, was christened the "Eranus," and three of Maxwell's contributions to it have been preserved and are printed by Professor Campbell.

After the Cavendish papers were finished, Maxwell had more time for his own original researches, and two important papers were published in 1879. The one on "Stresses in Rarefied Gases arising from Inequalities of Temperature" was printed in the Royal Society's Transactions, and deals with the Theory of the Radiometer; the other on "Boltzmann's Theorem" appears in the Transactions of the Cambridge Philosophical Society. In the previous year he had delivered the Rede lecture on "The Telephone." He also began to prepare a second edition of "Electricity and Magnetism."

His health gave way during the Easter term of 1879; indeed for two years previously he had been troubled with dyspeptic symptoms, but had consulted no one on the subject. He left Cambridge as usual in June, hoping that he would quickly recover at Glenlair, but he grew worse instead. In October he was told by Dr. Sanders of Edinburgh that he had not a month to live. He returned to Cambridge in order to be under the care of Dr. Paget, who was able in some measure to relieve his most severe suffering but the disease, of which his mother had died at the same age, continued its progress, and he died on November 5th. His one care during his last illness was for those whom he left behind. Mrs. Maxwell was an invalid dependent on him for everything, and the thought of her helplessness was the one thing which in these last days troubled him.

A funeral service took place in the chapel at Trinity College, and afterwards his remains were conveyed to Scotland and interred in the family burying-place at Corsock, Kirkcudbright.

A memorial edition of his works was issued by the Cambridge University Press in 1890. A portrait by Lowes Dickinson hangs in the hall of Trinity College, and there is a bust by Boehm in the laboratory.

After his death Mrs. Maxwell gave his scientific library to the Cavendish Laboratory, and on her death she left a sum of about £6,000 to found a scholarship in Physics, to be held at the laboratory.

The preceding pages contain some account of Clerk Maxwell's life as a man of science. His character had other sides, and any life of him would be incomplete without some brief reference to these. His letters to his wife and to other intimate friends show throughout his life the depth of his religious convictions. The high purpose evidenced in the paper given to the present Dean of Canterbury when leaving Cambridge, animated him continually, and appears from time to time in his writings. The student's evening hymn, composed in 1853 when still an undergraduate, expresses the same feelings—

> Through the creatures Thou hast made
>   Show the brightness of Thy glory,
> Be eternal truth displayed
>   In their substance transitory,
> Till green earth and ocean hoary,
>   Massy rock and tender blade,
> Tell the same unending story,
>   "We are Truth in form arrayed."
>
> Teach me so Thy works to read
>   That my faith, new strength accruing,
> May from world to world proceed,
>   Wisdom's fruitful search pursuing,
> Till Thy breath my mind imbuing,
>   I proclaim the eternal creed,
> Oft the glorious theme renewing,
>   God our Lord is God indeed.

His views on the relation of Science to Faith are given in his letter* to Bishop Ellicott already referred to—

"But I should be very sorry if an interpretation founded on a most conjectural scientific hypothesis were to get fastened to the text in Genesis, even if by so doing it got rid of the old statement of the commentators which has long ceased to be intelligible. The rate of change of scientific hypothesis is naturally much more rapid than that of Biblical interpretations, so that if an interpretation is founded on such an hypothesis, it may help to keep the hypothesis above ground long after it ought to be buried and forgotten.

"At the same time I think that each individual man should do all he can to impress his own mind with the extent, the order, and the unity of the universe, and should carry these ideas with him as he reads such passages as the 1st chapter of the Epistle to Colossians (*see* 'Lightfoot on Colossians,' p. 182), just as enlarged conceptions of the extent and unity of the world of life may be of service to us in reading Psalm viii., Heb. ii. 6, etc."

* "Life of J. C. Maxwell," p. 394.

And again in his letter* to the secretary of the Victoria Institute giving his reasons for declining to become a member—

"I think men of science as well as other men need to learn from Christ, and I think Christians whose minds are scientific are bound to study science, that their view of the glory of God may be as extensive as their being is capable of. But I think that the results which each man arrives at in his attempts to harmonise his science with his Christianity ought not to be regarded as having any significance except to the man himself, and to him only for a time, and should not receive the stamp of a society."

Professor Campbell and Mr. Garnett have given us the evidence of those who were with him in his last days, as to the strength of his own faith. On his death bed he said that he had been occupied in trying to gain truth; that it is but little of truth that man can acquire, but it is something to know in whom we have believed.

* "Life of J. C. Maxwell," p. 401.

## CHAPTER VII.

### SCIENTIFIC WORK—COLOUR VISION.

FIFTEEN years only have passed since the death of Clerk Maxwell, and it is almost too soon to hope to form a correct estimate of the value of his work and its relation to that of others who have laboured in the same field.

Thus Niven, at the close of his obituary notice in the Proceedings of the Royal Society, says: "It is seldom that the faculties of invention and exposition, the attachment to physical science and capability of developing it mathematically, have been found existing in one mind to the same degree. It would, however, require powers somewhat akin to Maxwell's own to describe the more delicate features of the works resulting from this combination, every one of which is stamped with the subtle but unmistakable impress of genius." And again in the preface to Maxwell's works, issued in 1890, he wrote: "Nor does it appear to the present editor that the time has yet arrived when the quickening influence of Maxwell's mind on modern scientific thought can be duly estimated."

It is, however, the object of the present series to attempt to give some account of the work of men of science of the last hundred years, and to show how each has contributed his share to our present stock of knowledge. This task, then, remains to be done.

While attempting it I wish to express my indebtedness to others who have already written about Maxwell's scientific work, especially to Mr. W. D. Niven, whose preface to the Maxwell papers has been so often referred to; to Mr. Garnett, the author of Part II. of the "Life of Maxwell," which deals with his contributions to science; and to Professor Tait, who in *Nature* for February 5th, 1880, gave an account of Clerk Maxwell's work, "necessarily brief, but sufficient to let even the non-mathematical reader see how very great were his contributions to modern science" —an account all the more interesting because, again to quote from Professor Tait, "I have been intimately acquainted with him since we were schoolboys together."

Maxwell's main contributions to science may be classified under three heads—"Colour Perception," "Molecular Physics," and "Electrical Theories." In addition to these there were other papers of the highest interest and importance, such as the essay on "Saturn's Rings," the paper on the "Equilibrium of Elastic Solids," and various memoirs on pure geometry and questions of mechanics, which would, if they stood alone, have secured for their author a distinguished position as a physicist and mathematician, but which are not the works by which his name will be mostly remembered.

The work on "Colour Perception" was begun at an early date. We have seen Maxwell while still at Edinburgh interested in the discussions about Hay's theories.

His first published paper on the subject was a

letter to Dr. G. Wilson, printed in the Transactions of the Royal Society of Arts for 1855; but he had been mixing colours by means of his top for some little time previously, and the results of these experiments are given in a paper entitled "Experiments on Colour," communicated to the Royal Society of Edinburgh by Dr. Gregory, and printed in their Transactions, vol. xxi.

In the paper on "The Theory of Compound Colours," printed in the Philosophical Transactions for 1860, Maxwell gives a history of the theory as it was known to him.

He points out first the distinction between the *optical* properties and the *chromatic* properties of a beam of light. "The optical properties are those which have reference to its origin and propagation through media until it falls on the sensitive organ of vision;" they depend on the periods and amplitudes of the ether vibrations which compose the beam. "The chromatic properties are those which have reference to its power of exciting certain sensations of colour perceived through the organ of vision." It is possible for two beams to be optically very different and chromatically alike. The converse is not true; two beams which are optically alike are also chromatically alike.

The foundation of the theory of compound colours was laid by Newton. He first shewed that "by the mixture of homogeneal light colours may be produced which are like to the colours of homogeneal light as to the appearance of colour, but not as to the immutability of colour and constitution of light." Two beams which differ optically may yet be alike chromatically; it is possible by mixing red and yellow to obtain an orange colour chromatically similar to the orange of the spectrum, but optically different to that orange, for the compound orange can be analysed by a prism into its component red and yellow; the spectrum orange is incapable of further resolution.

Newton also solves the following problem:—

*In a mixture of primary colours, the quantity and quality of each being given to know the colour of the compound* (Optics, Book 1, Part 2, Prop. 6), and his solution is the following:—He arranges the seven colours of the spectrum round the circumference of a circle, the length occupied by each colour being proportional to the musical interval to which, in Newton's views, the colour corresponded. At the centre of gravity of each of these arcs he supposes a weight placed proportional to the number of rays of the corresponding colour which enter into the mixture under consideration. The position of the centre of gravity of these weights indicates the nature of the resultant colour. A radius drawn through this centre of gravity points out the colour of the spectrum which it most resembles; the distance of the centre of gravity from the centre gives the fulness of the colour. The centre itself is white. Newton gives no proof of this rule; he merely says, "This rule I conceive to be accurate enough for practice, though not mathematically accurate."

Maxwell proved that Newton's method of finding the centre of gravity of the component colours was confirmed by his observations, and that it involves mathematically the theory of three elements of colour; but the disposition of the colours on the circle was only a provisional arrangement; the true relations of the colours could only be determined by direct experiment.

Thomas Young appears to have been the next, after Newton, to work at the theory of colour sensation. He made observations by spinning coloured discs much in the same way as that which was afterwards adopted by Maxwell, and he developed the theory that three different primary sensations may be excited in the eye by light, while the colour of any beam depends on the proportions in which these three sensations are excited. He supposes the three primary sensations to correspond to red, green, and violet. A blue ray is capable of exciting both the green and the violet; a yellow ray excites the red and the green. Any colour, according to Young's theory, may be matched by a mixture of these three primary colours taken in proper proportion; the quality of the colour depends on the proportion of the intensities of the components; its brightness depends on the sum of these intensities.

Maxwell's experiments were undertaken with the object of proving or disproving the physical part of Young's theory. He does not consider the question whether there are three distinct sensations corresponding to the three primary colours; that is a physiological inquiry, and one to which no completely satisfactory answer has yet been given. He does show that by a proper mixture of any three arbitrarily chosen standard colours it is possible to match any other colour; the words "proper mixture," however, need, as will appear shortly, some development.

We may with advantage compare the problem with one in acoustics.

When a compound musical note consisting of a pure tone and its overtones is sounded, the trained ear can distinguish the various overtones and analyse the sound into its simple components. The same sensation cannot be excited in two different ways. The eye has no such corresponding power. A given yellow may be a pure spectral yellow, corresponding to a pure tone in music, or it may be a mixture of a number of other pure tones; in either case it can be matched by a proper combination of three standard colours — this Maxwell proved. It may be, as Young supposed, that if the three standard colours be properly selected they correspond exactly to three primary sensations of the brain. Maxwell's experiments do not afford any light on this point, which still remains more than doubtful.

When Maxwell began his work the theory of colours was exciting considerable interest. Sir David Brewster had recently developed a new theory of colour sensation which had formed the basis of some discussions, and in 1852 von Helmholtz published his first paper on the subject. According to Brewster, the three primitive colours were red, yellow and blue, and he supposed that they corresponded to three different kinds of objective light. Helmholtz pointed out that experiments up to that date had been conducted by mixing pigments, with the exception of those in which the rotating disc was used, and that it is necessary to make them on the rays of the spectrum itself. He then describes a method of mixing the light from two spectra so as to obtain the combination of every two of the simple prismatic rays in all degrees of relative strength.

From these experiments results, which at the time were unexpected, but some of which must have been known to Young, were obtained. Among them it was shown that a mixture of red and green made yellow, while one of green and violet produced blue.

In a later paper (*Philosophical Magazine*, 1854) Helmholtz described a method for ascertaining the various pairs of complementary colours—colours, that is, which when mixed will give white—which had been shown by Grassman to exist if Newton's theory were true. He also gave a provisional diagram of the curve formed by the spectrum, which ought to take the place of the circle in Newton's diagram; for this, however, his experiments did not give the complete data.

Such was the state of the question when Maxwell began. His first colour-box was made in 1852. Others were designed in 1855 and 1856, and the final paper appeared in 1860. But before that time he had established important results by means of his rotatory discs and colour top. In his own description of this he says: "The coloured paper is cut into the form of disc, each with a hole in the centre and divided along a radius so as to admit of several of them being placed on the same axis, so that part of each is exposed. By slipping one disc over another we can expose any given portion of each colour.

These discs are placed on a top or teetotum, which is spun rapidly. The axis of the top passes through the centre of the discs, and the quantity of each colour exposed is measured by graduations on the rim of the top, which is divided into 100 parts. When the top is spun sufficiently rapidly, the impressions due to each colour separately follow each other in quick succession at each point of the retina, and are blended together; the strength of the impression due to each colour is, as can be shown experimentally, the same as when the three kinds of light in the same relative proportions enter the eye simultaneously. These relative proportions are measured by the areas of the various discs which are exposed. Two sets of discs of different radius are used; the largest discs are put on first, then the smaller, so that the centre portion of the top shows the colour arising from the mixture of those of the smaller discs; the outer portion, that of the larger discs."

In experimenting, six discs of each size are used, black, white, red, green, yellow and blue. It is found by experiment that a match can be arranged between any five of these. Thus three of the larger discs are placed on the top—say black, yellow and blue—and two of the smaller discs, red and green, are placed above these. Then it is found that it is possible so to adjust the amount exposed of each disc that the two parts of the tóp appear when it is spun to be of the same tint. In one series of experiments the chromatic effect of 46·8 parts of black, 29·1 of yellow, and 24·1 of blue was found to be the same as that of 66·6 of red and 33·4 of green; each set of discs has a dirty yellow tinge.

Now, in this experiment, black is not a colour; practically no light reaches the eye from a dead black. We have, however, to fill up the circumference of the top in some way which will not affect the impression on the retina arising from the mixture of the blue and yellow; this we can do by using the black disc.

Thus we have shown that 66·6 parts of red and 33·4 parts of green produce the same chromatic effect as 29·1 of yellow and 24·1 of blue. Similarly in this manner a match can be arranged between any four colours and black, the black being necessary to complete the circumference of the discs.

Thus using A, B, C, D to denote the various colours, $a, b, c, d$ the amounts of each colour taken, we can get a series of results expressed as follows: $a$ parts of A together with $b$ parts of B match $c$ parts of C together with $d$ parts of D; or we may write this as an equation thus:—

$$a A + b B = c C + d D.$$

where the + stands for "combined with," and the = for "matches in tint."

We may also write the above—

$$d D = a A + b B - c C.$$

or $d$ parts of D can be matched by a *proper* combination of colours A, B, C. The sign − shows that in order to make the match we have to combine the colour C with D; the combination then matches a mixture of A and B.

In this way we can form a number of equations for all possible colours, and if we like to take any three colours A, B, C as standards, we obtain a result which may be written generally—

$$x X = a A + b B + c C.$$

or $x$ parts of X can be matched by $a$ parts of A, combined with $b$ parts of B and $c$ parts of C. If the sign of one of the quantities $a$, $b$, or $c$ is negative, it indicates that that colour must be combined with X to match the other two.

Now Maxwell was able to show that, if A, B, C be properly selected, nearly every other colour can be matched by positive combinations of these three. These three colours, then, are primary colours, and nearly every other colour can be matched by a combination of the three primary colours.

Experiments, however, with coloured discs, such as were undertaken by Young, Forbes and Maxwell, were not capable of giving satisfactory results. The colours of the discs were not pure spectrum colours, and varied to some extent with the nature of the incident light. It was for this reason that Helmholtz in 1852 experimented with the spectrum, and that Maxwell about the same time invented his colour box.

The principle of the latter was very simple. Suppose we have a slit S, and some arrangement for forming a pure spectrum on a screen. Let there now be a slit R placed in the red part of the spectrum on the screen. When light falls on the slit S, only the red rays can reach R, and hence conversely, if the white source be placed at the other end of the apparatus, so that R is illuminated with white light, only red rays will reach S. Similarly, if another slit be placed in the green at G, and this be illuminated by white light, only the green rays will reach S, while from a third slit V in the violet, violet light only can arrive at S. Thus by opening the three slits at V, G and R simultaneously, and looking through S, the retina receives the impression of the three different colours. The amount of light of each colour will depend on the breadth to which the corresponding slit is opened, and the relative intensities of the three different components can be compared by comparing the breadths of the three slits. Any other colour which is allowed by some suitable contrivance to enter the eye simultaneously can now be matched, provided the red, green and violet are primary colours.

By means of experiments with the colour box Maxwell showed conclusively that a match could be obtained between any four colours; the experiments could not be carried out in quite the simple manner suggested by the above description of the principle of the box. An account of the method will be found in Maxwell's own paper. It consisted in matching a standard white by various combinations of other colours.

The main object of his research, however, was to examine the chromatic properties of the different parts of the spectrum, and to determine the form of the curve which ought to replace the circle in Newton's diagram of colour.

Maxwell adopted as his three standard colours: red, of about wave length 6,302; green, wave length 5,281; and violet, 4,569 tenth metres. On the scale of Maxwell's instrument these are represented by the numbers 24, 44 and 68.

Let us take three points A, B, C at the corners of an equilateral triangle to represent on a diagram these three colours. The position of any other colour on the diagram will be found by taking weights proportional to the amounts of the colours A, B, C required to make the match between A, B, C and the given colour; these weights are placed at A, B, C respectively; the position of their centre of gravity is the point required. Thus the position of white is given by the equation—

$$W = 18 \cdot 6\,(24) + 31 \cdot 4\,(44) + 30 \cdot 5\,(68)$$

which means that weights proportional to $18 \cdot 6$, $31 \cdot 4$ and $30 \cdot 5$ are to be placed at A, B, C respectively, and their centre of gravity is to be found. The point so found is the position of white. Any other colour is given by the equation—

$$X = a\,(24) + b\,(44) + c\,(68).$$

Again, the position on the diagram for all colours for which $a$, $b$, $c$ are all positive lies within the triangle A B C. If one of the co-efficients, say $c$, is negative the same construction applies, but the weight applied at C must be treated as acting in the opposite direction to those at A and B. A mixture of the given colour and C matches a mixture of A and B. It is clear that the point corresponding to X will then lie outside the triangle

A B C. Maxwell showed that, with his standards, nearly all colours could be represented by points inside the triangle. The colours he had selected as standards were very close to primary colours.

Again, he proved that any spectrum colour between red and green, when combined with a very slight admixture of violet, could be matched, in the case of either Mrs. Maxwell or himself, by a proper mixture of the red and green. The positions, therefore, of the spectrum colours between red and green lie just outside the triangle A B C, being very close to the line A B, while for the colours between green and violet Maxwell obtained a curve lying rather further outside the side B C. Any spectrum colour between green and violet, together with a slight admixture of red, can be matched by a proper mixture of green and violet.

Thus the circle of Newton's diagram should be replaced by a curve, which coincides very nearly with the two sides A B and B C of Maxwell's figure. Strictly, according to his observations, the curve lies just outside these two sides. The purples of the spectrum lie nearly along the third side, C A, of the triangle, being obtained approximately by mixing the violet and the red.

To find the point on the diagram corresponding to the colour obtained by mixing any two or more spectrum colours we must, in accordance with Newton's rule, place weights at the points corresponding to the selected colours, and find the centre of gravity of these weights.

This, then, was the outcome of Maxwell's work on

colour. It verified the essential part of Newton's construction, and obtained for the first time the true form of the spectrum curve on the diagram.

The form of this curve will of course depend on the eye of the individual observer. Thus Maxwell and Mrs. Maxwell both made observations, and distinct differences were found in their eyes. It appears, however, that a large majority of persons have normal vision, and that matches made by one such person are accepted by most others as satisfactory. Some people, however, are colour blind, and Maxwell examined a few such. In the case of those whom he examined it appeared as though vision was dichromatic, the red sensation seemed to be absent; nearly all colours could be matched by combinations of green and violet. The colour diagram was reduced to the straight line B C. Other forms of colour blindness have since been investigated.

In awarding to Maxwell the Rumford medal in 1860, Major-General Sabine, vice-president of the Royal Society, after explaining the theory of colour vision and the possible method of verifying it, said: "Professor Maxwell has subjected the theory to this verification, and thereby raised the composition of colours to the rank of a branch of mathematical physics," and he continues: "The researches for which the Rumford medal is awarded lead to the remarkable result that to a very near degree of approximation all the colours of the spectrum, and therefore all colours in nature which are only mixtures of these, can be perfectly imitated by mixtures of three actually attainable colours, which are the red, green and blue belonging respectively to three particular parts of the spectrum.

It should be noticed in concluding our remarks on this part of Maxwell's work that his results are purely physical. They are not inconsistent with the physiological part of Young's theory, viz., that there are three primary sensations of colour which can be transmitted to the brain, and that the colour of any object depends on the relative proportions in which these sensations are excited, but they do not prove that theory. Any physiological theory which can be accepted as true must explain Maxwell's observations, and Young's theory does this: but it is, of course, possible that other theories may explain them equally well, and be more in accordance with physiological observations than Young's. Maxwell has given us the physical facts which have to be explained: it is for the physiologists to do the rest.

## CHAPTER VIII.

### SCIENTIFIC WORK—MOLECULAR THEORY.

MAXWELL in his article "Atom," in the ninth edition of the *Encyclopædia Britannica*, has given some account of Modern Molecular Science, and in particular of the molecular theory of gases. Of this science, Clausius and Maxwell are the founders, though to their names it has recently been shown that a third, that of Waterston, must be added. In the present chapter it is intended to give an outline of Maxwell's contributions to molecular science, and to explain the advances due to him.

The doctrine that bodies are composed of small particles in rapid motion is very ancient. Democritus was its founder, Lucretius—de Rerum Naturâ—explained its principles. The atoms do not fill space: there is void between.

> "Quapropter locus est intactus inane vacansque,
> Quod si non esset, nullâ ratione moveri
> Res possent; namque officium quod corporis extat
> Officere atque obstare, id in omni tempore adesset
> Omnibus. Haud igitur quicquam procedere posset
> Principium quoniam cedendi nulla daret res."

According to Boscovitch an atom is an indivisible point, having position in space, capable of motion, and possessing mass. It is also endowed with the power of exerting force, so that two atoms attract or repel each other with a force depending on their distance apart. It has no parts or dimensions: it is a mere geometrical point without extension in space: it has not the property of impenetrability, for two atoms can, it is supposed, exist at the same point.

In modern molecular science according to Maxwell, "we begin by assuming that bodies are made up of parts each of which is capable of motion, and that these parts act on each other in a manner consistent with the principle of the conservation of energy. In making these assumptions we are justified by the facts that bodies may be divided into smaller parts, and that all bodies with which we are acquainted are conservative systems, which would not be the case unless their parts were also conservative systems.

"We may also assume that these small parts are in motion. This is the most general assumption we can make, for it includes as a particular case the theory that the small parts are at rest. The phenomena of the diffusion of gases and liquids through each other show that there may be a motion of the small parts of a body which is not perceptible to us.

"We make no assumption with respect to the nature of the small parts—whether they are all of one magnitude. We do not even assume them to have extension and figure. Each of them must be measured by its mass, and any two of them must, like visible bodies, have the power of acting on one another when they come near enough to do so. The properties of the body or medium are determined by the configuration of its parts."

These small particles are called molecules, and a

molecule in its physical aspect was defined by Maxwell in the following terms :—

"A molecule of a substance is a small body, such that if, on the one hand, a number of similar molecules were assembled together, they would form a mass of that substance; while on the other hand, if any portion of this molecule were removed, it would no longer be able, along with an assemblage of other molecules similarly treated, to make up a mass of the original substance."

We are to look upon a gas as an assemblage of molecules flying about in all directions. The path of any molecule is a straight line, except during the time when it is under the action of a neighbouring molecule; this time is usually small compared with that during which it is free.

The simplest theory we could formulate would be that the molecules behaved like elastic spheres, and that the action between any two was a collision following the laws which we know apply to the collision of elastic bodies. If the average distance between two molecules be great compared with their dimensions, the time during which any molecule is in collision will be small compared with the interval between the collisions, and this is in accordance with the fundamental assumption just mentioned. It is not, however, necessary to suppose an encounter between two molecules to be a collision. One molecule may act on another with a force, which depends on the distance between them, of such a character that the force is insensible except when the molecules are extremely close together.

It is not difficult to see how the pressure exerted

by a gas on the sides of a vessel which contains it may be accounted for on this assumption. Each molecule as it strikes the side has its momentum reversed—the molecules are here assumed to be perfectly elastic.

Thus each molecule of the gas is continually gaining momentum from the sides of the vessel, while it gives up to the vessel the momentum which it possessed before the impact. The rate at which this change of momentum proceeds across a given area measures the force exerted on that area; the pressure of the gas is the rate of change of momentum per unit of area of the surface.

Again, it can be shown that this pressure is proportional to the product of the mass of each molecule, the number of molecules in a unit of volume, and the square of the velocity of the molecules.

Let us consider in the first instance the case of a jet of sand or water of unit cross section which is playing against a surface. Suppose for the present that all the molecules which strike the surface have the same velocity.

Then the number of molecules which strike the surface per second, will be proportional to this velocity. If the particles are moving quickly they can reach the surface in one second from a greater distance than is possible if they be moving slowly. Again, the number reaching the surface will be proportional to the number of molecules per unit of volume. Hence, if we call $v$ the velocity of each particle, and $N$ the number of particles per unit of volume, the number which strike the surface in one second will be $N v$;

if $m$ be the mass of each molecule, the mass which strikes the surface per second is $N m v$; the velocity of each particle of this mass is $v$, therefore the momentum destroyed per second by the impact is $N m v \times v$, or $N m v^2$, and this measures the pressure.

Hence in this case if $p$ be the pressure

$$p = N m v^2.$$

In the above we assume that *all* the molecules in the jet are moving with velocity $v$ perpendicular to the surface. In the case of a crowd of molecules flying about in a closed space this is clearly not true. The molecules may strike the surface in any direction; they will not all be moving normal to the surface. To simplify the case, consider a cubical box filled with gas. The box has three pairs of equal faces at right angles. We may suppose one-third of the particles to be moving at right angles to each face, and in this case the number per unit volume which we have to consider is not $N$, but $\frac{1}{3} N$. Hence the formula becomes $p = \frac{1}{3} N m v^2$.

Moreover, if $\rho$ be the density of the gas—that is, the mass of unit volume—then $Nm$ is equal to $\rho$, for $m$ is the mass of each particle, and there are $N$ particles in a unit of volume.

Hence, finally, $p = \frac{1}{3} \rho v^2$.

Or, again, if $V$ be the volume of unit mass of the gas, then $\rho V$ is unity, or $\rho$ is equal to $1/V$.

Hence $pV = \frac{1}{3} v^2$.

Formulæ equivalent to these appear first to have been obtained by Herapath about the year 1816 (Thomson's "Annals of Philosophy," 1816). The results only, however, were stated in that year. A paper which attempted to establish them was presented to the Royal Society in 1820. It gave rise to very considerable correspondence, and was withdrawn by the author before being read. It is printed in full in Thomson's "Annals of Philosophy" for 1821, vol. i., pp. 273, 340, 401. The arguments of the author are no doubt open to criticism, and are in many points far from sound. Still, by considering the problem of the impact of a large number of hard bodies, he arrived at a formula connecting the pressure and volume of a given mass of gas equivalent to that just given. These results are contained in Propositions viii. and ix. of Herapath's paper.

In his next step, however, Herapath, as we know now, was wrong. One of his fundamental assumptions is that the temperature of a gas is measured by the momentum of each of its particles. Hence, assuming this, we have $T = m\,v$, if $T$ represents the temperature; and

$$p = \tfrac{1}{3} N\,m\,v^2 = \tfrac{1}{3}\frac{N}{m}(m\,v)^2.$$

Or, again—

$$p = \tfrac{1}{3} N \cdot T \cdot v = \tfrac{1}{3} \cdot \frac{N}{m} \cdot T^2.$$

These results are practically given in Proposition viii., Corr. (1) and (2), and Proposition ix.* The tempera-

---

* In his "Hydrodynamics," published in 1738, Daniel Bernouilli had discussed the constitution of a gas, and had proved from general considerations that the pressure, if it arose from the impact of a number of moving particles, must be proportional to the square of their velocity. (*See* "Pogg. Ann.," Bd. 107, 1859, p. 490.)

H

ture as thus defined by Herapath is an absolute temperature, and he calculates the absolute zero of temperature at which the gas would have no volume from the above results. The actual calculation is of course wrong, for, as we know now by experiment, the pressure is proportional to the temperature, and not to its square, as Herapath supposed. It will be seen, however, that Herapath's formula gives Boyle's law: for if the temperature is constant, the formula is equivalent to

$$p\,V = \text{a constant.}$$

Herapath somewhat extended his work in his "Mathematical Physics" published in 1847, and applied his principles to explain diffusion, the relation between specific heat and atomic weight, and other properties of bodies. He still, however, retained his erroneous supposition that temperature is to be measured by the momentum of the individual particles.

The next step in the theory was made by Waterston. His paper was read to the Royal Society on March 5th, 1846. It was most unfortunately committed to the Archives of the Society, and was only disinterred by Lord Rayleigh in 1892 and printed in the Transactions for that year.

In the account just given of the theory, it has been supposed that all the particles move with the same velocity. This is clearly not the case in a gas. If at starting all the particles had the same velocity, the collisions would change this state of affairs. Some particles will be moving quickly, some slowly. We may, however, still apply the theory by splitting up the particles into groups, and, supposing that each group has a constant velocity, the particles in this group will contribute to the pressure an amount—$p_1$—equal to $\frac{1}{3} N_1 m v_1^2$, where $v_1$ is the velocity of the group and $N_1$ the number of particles having that velocity. The whole pressure will be found by adding that due to the various groups, and will be given as before by $p = \frac{1}{3} N m v^2$, where $v$ is not now the actual velocity of the particles, but a mean velocity given by the equation

$$N v^2 = N_1 v_1^2 + N_2 v_2^2 + \ldots ,$$

which will produce the same pressure as arises from the actual impacts. This quantity $v^2$ is known as the *mean square* of the molecular velocity, and is so used by Waterston.

In a paper in the *Philosophical Magazine* for 1858 Waterston gives an account of his own paper of 1846 in the following terms:—"Mr. Herapath unfortunately assumed heat or temperature to be represented by the simple ratio of the velocity instead of the square of the velocity, being in this apparently led astray by the definition of motion generally received, and thus was baffled in his attempts to reconcile his theory with observation. If we make this change in Mr. Herapath's definition of heat or temperature—viz., that it is proportional to the vis-viva or square velocity of the moving particle, not to the momentum or simple ratio of the velocity—we can without much difficulty deduce not only the primary laws of elastic fluids, but also the other physical properties of gases enumerated above in the third objection to Newton's hypothesis. [The paper from which the quotation is taken is on 'The Theory of Sound.'] In the Archives of the Royal Society for 1845-46 there is a paper on 'The Physics of Media that consist of perfectly "Elastic Molecules in a State of Motion,"' which contains the synthetical reasoning on which the demonstration of these matters rests. . . . This theory does not take account of the size of the molecules. It assumes that no time is lost at the impact, and that if the impacts produce rotatory motion, the vis viva thus invested bears a constant ratio to the rectilineal vis viva, so as not to require separate consideration. It does, also, not take account of the probable internal motion of composite molecules; yet the results so closely accord with observation in every part of the subject as to leave no doubt that Mr. Herapath's idea of the physical constitution of gases approximates closely to the truth."

In his introduction to Waterston's paper (Phil. Trans., 1892) Lord Rayleigh writes:—"Impressed with the above passage, and with the general ingenuity and soundness of Waterston's views, I took the first opportunity of consulting the Archives, and saw at once that the memoir justified the large claims made for it, and that it marks an immense advance in the direction of the now generally received theory."

In the first section of the paper Waterston's great advance consisted in the statement that the mean square of the kinetic energy of each molecule measures the temperature.

According to this we are thus to put in the pressure equation—$\tfrac{1}{2} m v^2 = T$, the temperature, and we have at once—$pV = \tfrac{2}{3} N \cdot T$.

Now this equation expresses, as we know, the laws of Boyle and Gay Lussac.

The second section discusses the properties of media, consisting of two or more gases, and arrives at the result that "in mixed media the mean square molecular velocity is inversely proportional to the specific weights of the molecules." This was the great law rediscovered by Maxwell fifteen years later. With modern notation it may be put thus:—If $m_1$, $m_2$ be the masses of each molecule of two different sets of molecules mixed together, then, when a steady state has been reached, since the temperature is the same throughout, $m_1 v_1^2$ is equal to $m_2 v_2^2$. The average kinetic energy of each molecule is the same.

From this Avogadros' law follows at once—for if $p_1$, $p_2$ be the pressures, $N_1$, $N_2$ the numbers of molecules per unit volume—

$$p_1 = \tfrac{1}{3} N_1 m_1 v_1^2,$$
$$p_2 = \tfrac{1}{3} N_2 m_2 v_2^2.$$

Hence, if $p_1$ is equal to $p_2$, since $m_1 v_1^2$ is equal to $m_2 v_2^2$, we must have $N_1$ equal to $N_2$, or the number of molecules in equal volumes of two gases at the same pressure and temperature is the same. The proof of this proposition given by Waterston is not satisfactory. On this point, however, we shall have more to say. The third section of the paper deals with adiabatic expansion, and in it there is an error in calculation which prevented correct results from being attained.

At the meeting of the British Association at Ipswich, in 1851, a paper by J. J. Waterston of Bombay, on "The General Theory of Gases," was read. The following is an extract from the Proceedings:—

The author "conceives that the atoms of a gas, being perfectly elastic, are in continual motion in all directions, being constrained within a limited space by their collisions with each other, and with the particles of surrounding bodies.

"The vis viva of these motions in a given portion of a gas constitutes the quantity of heat contained in it.

"He shows that the result of this state of motion must be to give the gas an elasticity proportional to the mean square of the velocity of the molecular motions, and to the total mass of the atoms contained in unity of bulk" (unit of volume)—that is to say, to the density of the medium.

"The elasticity in a given gas is the measure of temperature. Equilibrium of pressure and heat between two gases takes place when the number of atoms in unit of volume is equal and the vis viva of each atom equal. Temperature, therefore, in all gases is proportional to the mass of one atom multiplied by the mean square of the velocity of the molecular motions, being measured from an absolute zero 491° below the zero of Fahrenheit's thermometer."

It appears, therefore, from these extracts that the discovery of the laws that temperature is measured by the mean kinetic energy of a single molecule, and that in a mixture of gases the mean kinetic energy of each molecule is the same for each gas, is due to Waterston. They were contained in his paper of 1846, and published by him in 1851. Both these papers, however, appear to have been unnoticed by all subsequent writers until 1892.

Meanwhile, in 1848, Joule's attention was called by his experiments to the question, and he saw that Herapath's result gave a means of calculating the mean velocity of the molecules of a gas. For according to the result given above, $p = \frac{1}{3} \rho v^2$: thus $v^2 = 3 p/\rho$, and $p$ and $\rho$ being known, we find $v^2$. Thus for hydrogen at freezing-point and atmospheric pressure Joule obtains for $v$ the value 6,055 feet per second, or, roughly, six times the velocity of sound in air.

Clausius was the next writer of importance on the subject. His first paper is in "Poggendorff's Annalen," vol. c., 1857, "On the Kind of Motion we call Heat." It gives an exposition of the theory, and establishes the fact that the kinetic energy of the translatory motion of a molecule does not represent the whole of the heat it contains. If we look upon a molecule as a small solid we must consider the energy it possesses in consequence of its rotation about its centre of gravity, as well as the energy due to the motion of translation of the whole.

Clausius' second paper appeared in 1859. In it he considers the average length of the path of a molecule during the interval between two collisions. He determines this path in terms of the average distance between the molecules and the distance between the centres of two molecules at the time when a collision is taking place.

These two papers appear to have attracted Maxwell's attention to the matter, and his first paper, entitled "Illustrations of the Dynamical Theory of Gases," was read to the British Association at Aberdeen and Oxford in 1859 and 1860, and appeared in the *Philosophical Magazine*, January and July, 1860.

In the introduction to this paper Maxwell points out, while there was then no means of measuring the quantities which occurred in Clausius' expression for the mean free path, "the phenomena of the internal friction of gases, the conduction of heat through a gas, and the diffusion of one gas through another, seem to indicate the possibility of determining accurately the mean length of path which a particle describes between two collisions. In order, therefore, to lay the foundation of such investigations on strict mechanical principles," he continues, "I shall demonstrate the laws of motion of an indefinite number of small, hard and perfectly elastic spheres acting on one another only during impact."

Maxwell then proceeds to consider in the first case the impact of two spheres.

But a gas consists of an indefinite number of molecules. Now it is impossible to deal with each molecule individually, to trace its history and follow its path. In order, therefore, to avoid this difficulty Maxwell introduced the statistical method of dealing with such problems, and this introduction is the first great step in molecular theory with which his name is connected.

He was led to this method by his investigation into the theory of Saturn's rings, which had been completed in 1856, and in which he had shown that the conditions of stability required the supposition that the rings are composed of an indefinite number of free particles revolving round the planet, with velocities depending on their distances from the centre. These particles may either be arranged in separate rings, or their motion may be such that they are continually coming into collision with each other.

As an example of the statistical method, let us consider a crowd of people moving along a street. Taken as a whole the crowd moves steadily forwards. Any individual in the crowd, however, is jostled backwards and forwards and from side to side; if a line were drawn across the street we should find people crossing it in both directions. In a considerable interval more people would cross it, going in the direction in which the crowd is moving, than in the other and the velocity of the crowd might be estimated by counting the number which crossed the line in a given interval. This velocity so found would differ greatly from the velocity of any individual, which might have any value within limits, and which is continually changing. If we knew the velocity of each individual and the number of individuals we could calculate the average velocity, and this would agree with the value found by counting the resultant number of people who cross the line in a given interval.

Again, the people in the crowd will naturally fall into groups according to their velocities. At any moment there will be a certain number of people whose velocities are all practically equal, or, to be more accurate, do not differ among themselves by more than some small quantity. The number of people at any moment in each of these groups will be very different. The number in any group, which has a velocity not differing greatly from the mean velocity of the whole, will be large; comparatively few will have either a very large or a very small velocity.

Again, at any moment, individuals are changing from one group to another; a man is brought to a stop by some obstruction, and his velocity is considerably altered—he passes from one group to a different one; but while this is so, if the mean velocity remains constant, and the size of the crowd be very great, the number of people at any moment in a given group remains unchanged. People pass from that group into others, but during any interval the same number pass back again into that group.

It is clear that if this condition is satisfied the distribution is a steady one, and the crowd will continue to move on with the same uniform mean velocity.

Now, Maxwell applies these considerations to a crowd of perfectly elastic spheres, moving anyhow in a closed space, acting upon each other only when in contact. He shows that they may be divided into groups according to their velocities, and that, when the steady state is reached, the number in each group will remain the same, although the individuals change. Moreover, it is shown that, if A and B represent any two groups, the state will only be steady when the numbers which pass from the group A to the group B are equal to the numbers which pass back from the group B to the group A. This condition, combined with the fact that the total kinetic energy of the motion remains unchanged, enables him to calculate the number of particles in any group in terms of the whole number of particles, the mean velocity, and the actual velocity of the group.

From this an accurate expression can be found for the pressure of the gas, and it is proved that the value found by others, on the assumption that all the particles were moving with a common velocity, is correct. Previous to this paper of Maxwell's it had been realised that the velocities could not be uniform throughout. There had been no attempt to determine the distribution of velocity, or to submit the problem to calculation, making allowance for the variations in velocity.

Maxwell's mathematical methods are, in their generality and elegance, far in advance of anything previously attempted in the subject.

So far it has been assumed that the particles in the vessel are all alike. Maxwell next takes the case of a mixture of two kinds of particles, and inquires what relation must exist between the average velocities of these different particles, in order that the state may be steady.

Now, it can be shown that when two elastic spheres impinge the effect of the impact is always such as to reduce the difference between their kinetic energies.

Hence, after a very large number of impacts the kinetic energies of the two balls must be the same: the steady state, then, will be reached when each ball has the same kinetic energy.

Thus if $m_1$, $m_2$ be the masses of the particles in

the two sets respectively, $v_1$, $v_2$ their mean velocities we must have finally—

$$\tfrac{1}{2} m_1 v_1^2 = \tfrac{1}{2} m_2 v_2^2$$

This is the second of the two great laws enunciated by Waterston in 1845 and 1851, but which, as we have seen, had remained unknown until 1859, when it was again given by Maxwell.

Now, when gases are mixed their temperatures become equal. Hence we conclude, in Maxwell's words, "that the physical condition which determines that the temperature of two gases shall be the same, is that the mean kinetic energy of agitation of the individual molecules of the two gases are equal."

Thus, as the result of Maxwell's more exact researches on the motion of a system of spherical particles, we find that we again can obtain the equations—

$$T = \tfrac{1}{2} m v^2$$
$$p = \tfrac{1}{3} N m v^2 = \tfrac{2}{3} N T = \tfrac{2}{3} \rho \frac{T}{m}$$

From these results we obtain as before the laws of Boyle, Charles and Avrogadro.

Again if $\sigma$ be the specific heat of the gas at constant volume, the quantity of heat required to raise a single molecule of mass $m$ one degree will be $\sigma m$.

Thus, when a molecule is heated, the kinetic energy must increase by this amount. But the increase of temperature, which in this case is $1^0$, is measured by the increase of kinetic energy of the single molecule. Hence the amount of heat required to raise the temperature of a single molecule of all gases 1° is the same. Thus the quantity $\sigma m$ is the same for all gases; or, in other words, the specific heat of a gas is inversely proportional to the mass of its individual molecules. The density of a gas—since the number of molecules per unit volume at a given pressure and temperature is the same for all gases—is also proportional to the mass of each individual molecule. Thus the specific heats of all gases are inversely proportional to their densities. This is the law discovered experimentally by Dulong and Petit to be approximately true for a large number of substances.

In the next part of the paper Maxwell proceeded to determine the average number of collisions in a given time, and hence, knowing the velocities, to determine, in terms of the size of the particles and their numbers, the mean free path of a particle; the result so found differed somewhat from that already obtained by Clausius.

Having done this he showed how, by means of experiments on the viscosity of gases, the length of the mean free path could be determined.

An illustration due to Professor Balfour Stewart will perhaps make this clear. Let us suppose we have two trains running with uniform speed in opposite directions on parallel lines, and, further, that the engines continue to work at the same rate, developing just sufficient energy to overcome the resistance of the line, etc., and to maintain the speed constant. Now suppose passengers commence to jump across from one train to the other. Each man carries with him his own momentum, which is in the opposite direction to that of the train into which he jumps; the result is that the momentum of each train is reduced by the process; the velocities of the two decrease; it appears as though a frictional force were acting between the two. Maxwell suggests that a similar process will account for the apparent viscosity of gases.

Consider two streams of gas, moving in opposite directions one over the other; it is found that in each case the layers of gas near the separating surface move more slowly than those in the interior of the streams; there is apparently a frictional force between the two streams along this surface, tending to reduce their relative velocity. Maxwell's explanation of this is that at the common surface particles from the one stream enter the other, and carry with them their own momentum; thus near this surface the momentum of each stream is reduced, just as the momentum of the trains is reduced by the people jumping across. Internal friction or viscosity is due to the diffusion of momentum across this common surface. The effect does not penetrate far into the gas, for the particles soon acquire the velocity of the stream to which they have come.

Now, the rate at which the momentum is diffused will measure the frictional force, and will depend on the mean free path of the particles. If this is considerable, so that on the average a particle can penetrate a considerable distance into the second gas before a collision takes place and its motion is changed, the viscosity will be considerable; if, on the other hand, the mean free path is small, the reverse will be true. Thus it is possible to obtain a relation between the mean free path and the coefficient of viscosity, and from this, if the coefficient of viscosity be known, a value for the mean free path can be found.

Maxwell, in the paper under discussion, was the first to do this, and, using a value found by Professor Stokes for the coefficient of viscosity, obtained as the length of the mean free path of molecules of air $\frac{1}{1700}$ of an inch, while the number of collisions per second experienced by each molecule is found to be about 8,077,200,000.

Moreover, it appeared from his theory that the coefficient of viscosity should be independent of the number of molecules of gas present, so that it is not altered by varying the density. This result Maxwell characterises as startling, and he instituted an elaborate series of experiments a few years later with a view of testing it. The reason for this result will appear if we remember that, when the density is decreased, the mean free path is increased; relatively, then, to the total number of molecules present, the number which cross the surface in a given time is increased. And it appears from Maxwell's result that this relative increase is such that the total number crossing remains unchanged. Hence the momentum conveyed across each unit area per second remains the same, in spite of the decrease in density.

Another consequence of the same investigation is that the coefficient of viscosity is proportional to the mean velocity of the molecules. Since the absolute temperature is proportional to the square of the velocity, it follows that the coefficient of viscosity is proportional to the square root of the absolute temperature.

The second part of the paper deals with the process of diffusion of two or more kinds of moving particles among one another.

If two different gases are placed in two vessels separated by a porous diaphragm such as a piece of unglazed earthenware, or connected by means of a narrow tube, Graham had shewn that, after sufficient time has elapsed, the two are mixed together. The same process takes place when two gases of different density are placed together in the same vessel. At first the denser gas may be at the bottom, the less dense above, but after a time the two are found to be uniformly distributed throughout.

Maxwell attempted to calculate from his theory the rate at which the diffusion takes place in these cases. The conditions of most of Graham's experiments were too complicated to admit of direct comparison with the theory, from which it appeared that there is a relation between the mean free path and the rate of diffusion. One experiment, however, was found, the conditions of which could be made the subject of calculation, and from it Maxwell obtained as the value of the mean free path in air $\frac{1}{389000}$ of an inch.

The number was close enough to that found from the viscosity to afford some confirmation of his theory.

However, a few years later Clausius criticised the details of this part of the paper, and Maxwell, in his memoir of 1866, admits the calculation to have been erroneous. The main principles remained unaffected, the molecules pass from one gas to the other, and this constitutes diffusion.

Now, suppose we have two sets of particles in contact of such a nature that the mean kinetic energy of the one set is different from that of the other; the temperatures of the two will then be different. These two sets will diffuse into each other, and the diffusing particles will carry with them their kinetic energy, which will gradually pass from those which have the greater energy to those which have the less, until the average kinetic energy is equalised throughout. But the kinetic energy of translation is the heat of the particles. This diffusion of kinetic energy is a diffusion of heat by conduction, and we have here the mechanical theory of the conduction of heat in a gas.

Maxwell obtained an expression, which, however, he afterwards modified, for the conductivity of a gas in terms of the mean free path. It followed from this that the conductivity of air was only about $\frac{1}{...}$ of that of copper.

Thus the diffusion of gases, the viscosity of gases, and the conduction of heat in gases, are all connected with the diffusion of the particles carrying with them their momenta and their energy; while values of the mean free path can be obtained from observations on any one of these properties.

In the third part of his paper Maxwell considers

the consequences of supposing the particles not to be spherical. In this case the impacts would tend to set up a motion of rotation in the particles. The direction of the force acting on any particle at impact would not necessarily pass through its centre; thus by impact the velocity of its centre would be changed, and in addition the particles would be made to spin. Some part, therefore, of the energy of the particles will appear in the form of the translational energy of their centres, while the rest will take the form of rotational energy of each particle about its centre.

It follows from Maxwell's work that for each particle the average value of these two portions of energy would be equal. The total energy will be half translational and half rotational.

This theorem, in a more general form which was afterwards given to it, has led to much discussion, and will be again considered later. For the present we will assume it to be true. Clausius had already called attention to the fact that some of the energy must be rotational unless the molecules be smooth spheres, and had given some reasons for supposing that the ratio of the whole energy to the energy of translation is in a steady state a constant. Maxwell shows that for rigid bodies this constant is 2. Let us denote it for the present by the symbol $\beta$. Thus, if the translational energy of a molecule is $\frac{1}{2} m v^2$, its whole energy is $\frac{1}{2} \beta m v^2$.

The temperature is still measured by the translational energy, or $\frac{1}{2} m v^2$; the heat depends on the whole energy. Hence if H represent the amount of heat—measured as energy—contained by a single molecule, and T its temperature, we have—

$$H = \beta T$$

From this it can be shewn\* that if $\gamma$ represent the ratio of the specific heat of a gas at constant pressure to the specific heat at constant volume, then—

$$\beta = \frac{2}{3} \frac{1}{\gamma - 1}$$

For air and some other gases the value of $\gamma$ has been shown to be 1·408. From this it follows that

---

\* The proof is as follows :—

If $\sigma$ be the specific heat at constant volume, $\sigma'$ at constant pressure, and consider a unit of mass of gas at pressure p and volume v, let the volume increase by an amount dv, while the temperature dy.

Thus $\quad \sigma' dT = \sigma dT + p dv$

But $\quad pv = \frac{2}{3} \frac{T}{m}$

Hence p being constant,

$$p dv = \frac{2}{3} \frac{dT}{m}$$

Therefore $\quad \sigma' = \sigma + \frac{2}{3} \frac{1}{m}$

Now suppose an amount of heat, dH, is given to a single molecule and that its temperature is T. Its specific heat is $\sigma$, and

$$dH = \sigma m dT$$

But $\quad dH = \beta dT$

Therefore $\quad \beta = \sigma m$

Hence $\quad \frac{1}{m} = \frac{\sigma}{\beta}$

Thus $\quad \sigma' = \sigma \left(1 + \frac{2}{3\beta}\right)$

And $\quad \sigma'/\sigma = \gamma$

Therefore $\quad \gamma = 1 + \frac{2}{3\beta}$

Or $\quad \beta = \frac{2}{3(\gamma - 1)}$

$\beta = 1\cdot 634$. Now, Maxwell's theory required that for smooth hard particles, approximately spherical in shape, $\beta$ should be 2, and hence he concludes "we have shown that a system of such particles could not possibly satisfy the known relation between the two specific heats of all gases."

Since this statement was made many more experiments on the value of $\gamma$ have been undertaken; it is not equal to $1\cdot 408$ for *all* gases. Hence the value of $\beta$ is different for various gases.

It is of some importance to notice that the value of $\beta$ just found for air is very approximately $1\cdot 66$ or $\frac{5}{3}$.

For mercury vapour the value of $\gamma$ has been shown by Kundt to be $1\cdot 33$ or $1\frac{1}{3}$, and hence $\beta$ is equal to 1. Thus all the energy of a particle of mercury vapour is translational, and its behaviour in this respect is consistent with the assumption that a particle of mercury vapour is a smooth sphere.

The two results of this theory which seemed to lend themselves most readily to experimental verification were (1) that the viscosity of a gas is independent of its density, and (2) that it is proportional to the square root of the absolute temperature. The next piece of work connected with the theory was an attempt to test these consequences, and a description of the experiments was published in the "Philosophical Transactions" for 1865, in a paper on the "Viscosity or Internal Friction of Air and other Gases," and forms the Bakerian lecture for that year.

The first result was completely proved. It is

shewn that the value of the coefficient* of viscosity "is the same for air at 0·5 inch and at 30 inches pressure, provided that the temperature remains the same."

It was clear also that the viscosity depended on the temperature, and the results of the experiments seemed to show that it was nearly proportional to the absolute temperature. Thus for two temperatures, 185° Fah. and 51° Fah., the ratio of the two coefficients found was 1·2624; the ratio of the two temperatures, each measured from absolute zero, is 1·2605.

This result, then, does not agree with the hypothesis that a gas consists of spherical molecules acting only on each other by a kind of impact, for, if this were so, the coefficient would, as we have seen, depend on the square root of the absolute temperature. But Maxwell's result, connecting viscosity with the first power of the absolute temperature, has not been confirmed by other investigators. According to it we should have as the relation between $\mu$, the coefficient of viscosity at $t°$ and $\mu_0$, that at zero the equation—

$$\mu = \mu_0 (1 + .00365\, t).$$

The most recent results of Professor Holman (*Philosophical Magazine*, Vol. xxi., p. 212) give—

$$\mu = \mu_0 \{1 + .00275\, t - .00000034\, t^2\}.$$

And results similar to this are given by O. E. Meyer,

* Owing to an error of calculation the actual value obtained by Maxwell from these observations for the coefficient of viscosity is too great. More recent observers have found lower values than those given by him; the difference is thus explained.

Puluj, and Obermeyer. Maxwell's coefficient ·00365 is too large, but ·00182, the coefficient obtained by supposing the viscosity proportional to the square root of the temperature, would be too small.

It still remains true, therefore, that the laws of the viscosity of gases cannot be explained by the hypothesis of the impact of hard spheres; but some deductions drawn by Maxwell in his next paper from his supposed law of proportionality to the first power of the absolute temperature require modification.

It was clear from his experiments just described that the simple hypothesis of the impact of elastic bodies would not account for all the phenomena observed. Accordingly, in 1866, Maxwell took up the problem in a more general form in his paper on the "Dynamical Theory of Gases," Phil. Trans., 1866.

In it he considered the molecules of the gas not as elastic spheres of definite radius, but as small bodies, or groups of smaller molecules, repelling one another with a force whose direction always passes very nearly through the centre of gravity of the molecules, and whose magnitude is represented very nearly by some function of the distance of the centres of gravity. "I have made," he continues, "this modification of the theory in consequence of the results of my experiments on the viscosity of air at different temperatures, and I have deduced from these experiments that the repulsion is inversely as the fifth power of the distance."

Since more recent observation has shown that the numerical results of Maxwell's work connecting viscosity and temperature are erroneous, this last deduction does not hold; the inverse fifth power law of force will not give the correct relation between viscosity and temperature. Maxwell himself at a later date, "On the Stresses in Rarefied Gases," Phil. Trans., 1879, realised this; but even in this last paper he adhered to the fifth power law because it leads to an important simplification in the equations to be dealt with.

The paper of 1866 is chiefly important because it contains for the first time the application of general dynamical methods to molecular problems. The law of the distribution of velocities among the molecules is again investigated, and a result practically identical with that found for the elastic spheres is arrived at. In obtaining this conclusion, however, it is assumed that the distribution of velocities is uniform in all directions about any point, whatever actions may be taking place in the gas. If, for example, the temperature is different at different points, then, for a given velocity, all directions are not equally probable. Maxwell's expression, therefore, for the number of molecules which at any moment have a given velocity only applies to the permanent state in which the distribution of temperature is uniform. When dealing, for example, with the conduction of heat, a modification of the expression is necessary. This was pointed out by Boltzmann.*

In the paper of 1866, Maxwell applies his generalised results to the final distribution of two gases

* Studien über das Gleichgewicht der lebendigen Kraft zwischen bewegten materiellen Punkten Sitz d. k. Akad Wien, Band LVIII., 1868.

under the action of gravity, the equilibrium of temperature between two gases, and the distribution of temperature in a vertical column. These results are, as he states, independent of the law of force between the molecules. The dynamical causes of diffusion viscosity and conduction of heat are dealt with, and these involve the law of force.

It follows also from the investigation that, on the hypotheses assumed as its basis, if two kinds of gases be mixed, the difference between the average kinetic energies of translation of the gases of each kind diminishes rapidly in consequence of the action between the two. The average kinetic energy of translation, therefore, tends to become the same for each kind of gas, and as before, it is this average energy of translation which measures the temperature.

A molecule in the theory is a portion of a gas which moves about as a single body. It may be a mere point, a centre of force having inertia, capable of doing work while losing velocity. There may be also in each molecule systems of several such centres of force bound together by their mutual actions. Again, a molecule may be a small solid body of determinate form; but in this case we must, as Maxwell points out, introduce a new set of forces binding together the parts of each molecule: we must have a molecular theory of the second order. In any case, the most general supposition made is that a molecule consists of a series of parts which stick together, but are capable of relative motion among each other.

In this case the kinetic energy of the molecule consists of the energy of its centre of gravity, together with the energy of its component parts, relative to its centre of gravity.*

Now Clausius had, as we have seen, given reasons for believing that the ratio of the whole energy of a molecule to the energy of translation of its centre of gravity tends to become constant. We have already used $\beta$ to denote this constant. Thus, while the temperature is measured by the average kinetic energy of translation of the centre of gravity of each molecule, the heat contained in a molecule is its whole energy, and is $\beta$ times this quantity. Thus the conclusions as to specific heat, etc., already given on page 130, apply in this case, and in particular we have the result that if $\gamma$ be the ratio of the specific heat at constant pressure to that at constant volume, then—

$$\beta = \frac{2}{3}\frac{1}{\gamma - 1}$$

Maxwell's theorem of the distribution of kinetic energy among a system of molecules applied, as he gave it in 1866, to the kinetic energy of translation of the centre of gravity of each molecule. Two years later Dr. Boltzmann, in the paper we have already

* Another supposition which might be made, and which is necessary in order to explain various actions observed in a compound gas under electric force, is that the parts of which a molecule is composed are continually changing. Thus a molecule of steam consists of two parts of hydrogen, one of oxygen, but a given molecule of oxygen is not always combined with the same two molecules of hydrogen; the particles are continually changed. In Maxwell's paper an hypothesis of this kind is not dealt with.

referred to, extended it (under certain limitations) to the parts of which a molecule is composed. According to Maxwell the average kinetic energy of the centre of gravity of each molecule tends to become the same. According to Boltzmann the average kinetic energy of each part of the molecule tends to become the same.

Maxwell, in the last paper he wrote on the subject ("On Boltzmann's Theorem on the Average Distribution of Energy in a System of Material Points," Camb. Phil. Trans., XII.), took up this problem. Watson had given a proof of it in 1876 differing from Boltzmann's, but still limited by the stipulation that the time, during which a particle is encountering other particles, is very small compared with the time during which there is no sensible action between it and other particles, and also that the time during which a particle is simultaneously within the distance of more than one other particle may be neglected.

Maxwell claims that his proof is free from any such limitation. The material points may act on each other at all distances, and according to any law which is consistent with the conservation of energy; they may also be acted on by forces external to the system, provided these are consistent with that law.

The only assumption which is necessary for the direct proof is that the system, if left to itself in its actual state of motion, will sooner or later pass through every phase which is consistent with the conservation of energy.

In this paper Maxwell finds in a very general manner an expression for the number of molecules which at any time have a given velocity, and this, when simplified by the assumptions of the former papers, reduces to the form already found. He also shows that the average kinetic energy corresponding to any one of the variables which define his system is the same for every one of the variables of his system.

Thus, according to this theorem, if each molecule be a single small solid body, six variables will be required to determine the position of each, three variables will give us the position of the centre of gravity of the molecule, while three others will determine the position of the body relative to its centre of gravity. If the six variables be properly chosen, the kinetic energy can be expressed as a sum of six squares, one square corresponding to each variable. According to the theorem the part of the kinetic energy depending on each square is the same. Thus, the whole energy is six times as great as that which arises from any one of the variables. The kinetic energy of translation is three times as great as that arising from each variable, for it involves the three variables which determine the position of the centre of gravity. Hence, if we denote by K the kinetic energy due to one variable, the whole energy is 6 K, and the translational energy is 3 K; thus, for this case—

$$\beta = \frac{6\,K}{3\,K} = 2$$

Or, again, if we suppose that the molecule is such that $m$ variables are required to determine its position relatively to its centre of gravity, since 3 are needed to fix the centre of gravity, the total number

of variables defining the position of the molecule is
$m+3$, and it is said to have $m+3$ degrees of freedom.
Hence, in this case, its total energy is $(m+3)$ K and
its energy of translation is 3 K, thus we find—

$$\beta = \frac{m+3}{3}$$

Hence $\quad \gamma = 1 + \dfrac{2}{m+3} = 1 + \dfrac{2}{n}$

if $n$ be the number of degrees of freedom of the
molecule.

Thus, if this Boltzmann-Maxwell theorem be true,
the specific heat of a gas will depend solely on the
number of degrees of freedom of each of its molecules.
For hard rigid bodies we should have $n$ equal to 6,
and hence $\gamma = 1\cdot 333$. Now the fact that this is not
the value of $\gamma$ for any of the known gases is a
fundamental difficulty in the way of accepting the
complete theory.

Boltzmann has called attention to the fact that if
$n$ be equal to five, then $\gamma$ has the value $1\cdot 40$. And this
agrees fairly with the value found by experiment for
air, oxygen, nitrogen, and various other gases. We
will, however, return to this point shortly.

There is, perhaps, no result in the domain of
physical science in recent years which has been more
discussed than the two fundamental theorems of the
molecular theory which we owe to Maxwell and to
Boltzmann.

The two results in question are (1) the expression
for the number of molecules which at any moment
will have a given velocity, and (2) the proposition

that the kinetic energy is ultimately equally divided among all the variables which determine the system.

With regard to (1) Maxwell showed that his error law was one possible condition of permanence. If at any moment the velocities are distributed according to the error law, that distribution will be a permanent one. He did not prove that such a distribution is the only one which can satisfy all the conditions of the problem.

The proof that this law is a necessary, as well as a sufficient, condition of permanence was first given by Boltzmann, for a single monatomic gas in 1872, for a mixture of such gases in 1886, and for a polyatomic gas in 1887. Other proofs have been given since by Watson and Burbury. It would be quite beyond the limits of this book to go into the question of the completeness or sufficiency of the proofs. The discussion of the question is still in progress.

The British Association Report for 1894 contains an important contribution to the question, in the shape of a report by Mr. G. H. Bryan, and the discussion he started at Oxford by reading this report has been continued in the pages of *Nature* and elsewhere since that time.

Mr. Bryan shows in the first place what may be the nature of the systems of molecules to which the results will apply, and discusses various points of difficulty in the proof.

The theorem in question, from which the result (1) follows as a simple deduction, has been thus stated by Dr. Larmor.*

* *Nature*, vol. l., p. 152 (December 13th, 1894).

"There exists a positive function belonging to a group of molecules which, as they settle themselves into a steady state—on the average derived from a great number of configurations—maintains a steady downward trend. The Maxwell-Boltzmann steady state is the one in which this function has finally attained its minimum value, and is thus a unique steady state, it still being borne in mind that this is only a proposition of averages derived from a great number of instances in which nothing is conserved in encounters, except the energy, and that exceptional circumstances may exist, comparatively very few in number, in which the trend is, at any rate, temporarily the other way."

This theorem, when applied to cases of motion, such as that of a gas at constant temperature enclosed in a rigid envelope impermeable to heat, appears to be proved. For such a case, therefore, the Maxwell-Boltzmann law is the only one possible.

But whether this be so or not, the law first introduced by Maxwell is one of those possible, and the advance in molecular science due to its introduction is enormous.

We come now to the second result, the equal partition of the energy among all the degrees of freedom of each molecule. Lord Kelvin has pointed out a flaw in Maxwell's proof, but Boltzmann showed (*Philosophical Magazine*, March, 1893) how this flaw can easily be corrected, and it may be said that in all cases in which the Boltzmann-Maxwell law of the distribution of velocities holds, Maxwell's law of the equal partition of energy holds also.

Three cases are considered by Mr. Bryan, in which the law of distribution fails for rigid molecules: the first is when the molecules have all, in addition to their velocities of agitation, a common velocity of translation in a fixed direction; the second is when the gas has a motion of uniform rotation about a fixed axis; while the third is when each molecule has an axis of symmetry. In this last case the forces acting during a collision necessarily pass through the axis of symmetry, the angular velocity, therefore, of any molecule about this axis remains constant, the number of molecules having a given angular velocity will remain the same throughout the motion, and the part of the kinetic energy which depends on this component of the motion will remain fixed, and will not come into consideration when dealing with the equal partition of the energy among the various degrees of freedom.

Such a molecule has five, and not six, degrees of freedom; three quantities are needed to determine the position of its centre of gravity, and two to fix the position of the axis of symmetry.

In this case, then, as Boltzmann points out, in the expression for the ratio of the specific heats, we must have $n$ equal to 5, and hence

$$\gamma = 1 + \frac{2}{n} = 1 + \frac{2}{5} = 1\cdot4$$

agreeing fairly with the value found for air and various other permanent gases.

For cases, then, in which we consider each atom as a single rigid body, the Boltzmann-Maxwell theorem appears to give a unique solution, and the Maxwell law of the distribution of the energy to be in fair accordance with the results of observation.*

If we can never go further—and it must be admitted that the difficulties in the way of further advance are enormous—it may, I think, be claimed for Maxwell that the progress already made is greatly due to him. Both these laws, for the case of elastic spheres, are contained in his first paper of 1860; and while it is to the genius of Boltzmann that we owe their earliest generalisation, and in particular the proof of the uniqueness of the solution under proper restrictions, Maxwell's last paper contributed in no small degree to the security of the position. Not merely the foundations, but much of the superstructure of molecular science is his work.

The difficulties in the way of advance are, as we have said, enormous. Boltzmann, in one of his papers, has considered the properties of a complex molecule of a gas, consisting maybe of a number of atoms and possibly of ether atoms bound with them, and he concludes that such a molecule will behave in its progressive motion, and in its collisions with other molecules, nearly like a rigid body. But to quote from Mr. Bryan: "The case of a polyatomic molecule, whose atoms are capable of vibrating relative to one another, affords an interesting field for investigation and speculation. Is the Boltzmann distribution still unique, or do other permanent distributions exist in which the kinetic energy is unequally divided?"

* See papers by Mr. Capstick, *Phil. Trans*, vols. 185-186.

Again, the spectroscope reveals to us vibrations of the ether, which are connected in some way with the vibrations of the molecules of gas, whose spectrum we are observing. It seems clear that the law of equal partition does not apply to these, and yet, if we are to suppose that the ether vibrations are due to actual vibrations of the atoms which constitute a molecule, why does it not apply? Where does the condition come in which leads to failure in the proof? Or, again, is it, as has been suggested, the fact that the complex spectrum of a gas represents the terms of a Fourier Series, into which some elaborate vibration of the atoms is resolved by the ether? or is the spectrum due simply to electromagnetic vibrations on the surface of the molecules—vibrations whose period is determined chiefly by the size and shape of the molecule, but in which the atoms of which it is composed take part? There are grave difficulties in the way of either of these explanations, but we must not let our dread of the task which remains to be done blind our eyes to the greatness of Maxwell's work.

One other important paper, and a number of shorter articles, remain to be mentioned.

The Boltzmann-Maxwell law applies only to cases in which the temperature is uniform throughout. In a paper published in the Philosophical Transactions for 1879, on "Stresses in Rarefied Gases Arising from Inequalities of Temperature," Maxwell deals, among other matters, with the theory of the radiometer. He shows that the observed motions will not take place unless gas, in contact with a solid, can slide along the surface of the solid with a finite velocity between places where the temperature is different; and in an appendix he proves that, on certain assumptions regarding the nature of the contact of the solid and the gas, there will be, even when the pressure is constant, a flow of gas along the surface from the colder to the hotter parts.

Among his less important papers bearing on molecular theory must be mentioned a lecture on "Molecules" to the British Association at its Bradford meeting; "Scientific Papers of Clerk Maxwell," vol. ii., p. 361; and another on "The Molecular Constitution of Bodies," Scientific Papers, vol. ii., p. 418.

In this latter, and also in a review in *Nature* of Van der Waal's book on "The Continuity of the Gaseous and Liquid States,"* he explains and discusses Clausius' virial equation, by means of which the variations of the permanent gases from Boyle's law are explained. The lecture gives a clear account, in Maxwell's own inimitable style, of the advances made in the kinetic theory up to the date at which it was delivered, and puts clearly the difficulties it has to meet. Maxwell thought that those arising from the known values of the ratio of the specific heats were the most serious.

In the articles, "Atomic Constitution of Bodies" and "Diffusion," in the ninth edition of the *Encyclopædia Britannica*, we have Maxwell's later views on the fundamental assumptions of the molecular theory.

The text-book on "Heat" contains some further developments of the theory. In particular he shows

* *Nature*, vol. x.

how the conclusions of the second law of thermo-dynamics are connected with the fact that the coarseness of our faculties will not allow us to grapple with individual molecules.

The work described in the foregoing chapters would have been sufficient to secure to Maxwell a distinguished place among those who have advanced our knowledge; it remains still to describe his greatest work, his theory of Electricity and Magnetism.

## CHAPTER IX.

### SCIENTIFIC WORK.—ELECTRICAL THEORIES.

CLERK MAXWELL'S first electrical paper—that on Faraday's "Lines of Force"—was read to the Cambridge Philosophical Society on December 10th, 1855, and Part II. on February 11th, 1856. The author was then a Bachelor of Arts, only twenty-three years in age, and of less than one year's standing from the time of taking his degree.

The opening words of the paper are as follows (Scientific Papers, vol. i., p. 155):—

"The present state of electrical science seems peculiarly unfavourable to speculation. The laws of the distribution of electricity on the surface of conductors have been analytically deduced from experiment; some parts of the mathematical theory of magnetism are established, while in other parts the experimental data are wanting; the theory of the conduction of galvanism, and that of the mutual attraction of conductors, have been reduced to mathematical formulæ, but have not fallen into relation with the other parts of the science. No electrical theory can now be put forth, unless it shows the connection, not only between electricity at rest and current electricity, but between the attractions and inductive effects of electricity in both states. Such a theory must accurately satisfy those laws, the mathematical form of which is known, and must afford the means of calculating the effects in the limiting cases where the known formulæ are inapplicable. In order, therefore, to appreciate the requirements of the science, the student must make himself familiar with a considerable body of most intricate mathematics, the mere retention of which in the memory materially interferes with further progress. The first process, therefore, in the effectual study of the science, must be one of simplification and reduction of the results of previous investigation to a form in which the mind can grasp them. The results of this simplification may take the form of a purely mathematical formula or of a physical hypothesis. In the first case we entirely lose sight of the phenomena to be explained; and though we may trace out the consequences of given laws, we can never obtain more extended views of the connections of the subject. If, on the other hand, we adopt a physical hypothesis, we see the phenomena only through a medium, and are liable to that blindness to facts and rashness in assumption which a partial explanation encourages. We must therefore discover some method of investigation which allows the mind at every step to lay hold of a clear physical conception, without being committed to any theory founded on the physical science from which that conception is borrowed, so that it is neither drawn aside from the subject in pursuit of analytical subtleties, nor carried beyond the truth by a favourite hypothesis.

"In order to obtain physical ideas without adopting a physical theory we must make ourselves familiar with the existence of physical analogies. By a physical analogy I mean that partial similarity between the laws of one science and those of another which makes each of them illustrate the other. Thus all the mathematical sciences are founded on relations between physical laws and laws of numbers, so that the aim of exact science is to reduce the problems of Nature to the determination of quantities by operations with members. Passing from the most universal of all analogies to a very partial one, we find the same resemblance in mathematical form between two different phenomena giving rise to a physical theory of light.

"The changes of direction which light undergoes in passing from one medium to another are identical with the deviations of the path of a particle in moving through a narrow space in which intense forces act. This analogy, which extends only to the direction, and not to the velocity of motion, was long believed to be the true explanation of the refraction of light;

and we still find it useful in the solution of certain problems, in which we employ it without danger as an artificial method. The other analogy, between light and the vibrations of an elastic medium, extends much farther, but, though its importance and fruitfulness cannot be over-estimated, we must recollect that it is founded only on a resemblance *in form* between the laws of light and those of vibrations. By stripping it of its physical dress and reducing it to a theory of 'transverse alternations,' we might obtain a system of truth strictly founded on observation, but probably deficient both in the vividness of its conceptions and the fertility of its method. I have said thus much on the disputed questions of optics, as a preparation for the discussion of the almost universally admitted theory of attraction at a distance.

"We have all acquired the mathematical conception of these attractions. We can reason about them and determine their appropriate forms or formulæ. These formulæ have a distinct mathematical significance, and their results are found to be in accordance with natural phenomena. There is no formula in applied mathematics more consistent with Nature than the formula of attractions, and no theory better established in the minds of men than that of the action of bodies on one another at a distance. The laws of the conduction of heat in uniform media appear at first sight among the most different in their physical relations from those relating to attractions. The quantities which enter into them are *temperature, flow of heat, conductivity*. The word *force* is foreign to the subject. Yet we find that the mathematical laws of the uniform motion of heat in homogeneous media are identical in form with those of attractions varying inversely as the square of the distance. We have only to substitute *source of heat* for *centre of attraction*, *flow of heat* for *accelerating effect of attraction* at any point, and *temperature* for *potential*, and the solution of a problem in attractions is transformed into that of a problem in heat.

"This analogy between the formulæ of heat and attraction was, I believe, first pointed out by Professor William Thomson in the *Cambridge Mathematical Journal*, Vol. III.

"Now the conduction of heat is supposed to proceed by an

action between contiguous parts of a medium, while the force of attraction is a relation between distant bodies, and yet, if we knew nothing more than is expressed in the mathematical formulæ, there would be nothing to distinguish between the one set of phenomena and the other.

"It is true that, if we introduce other considerations and observe additional facts, the two subjects will assume very different aspects, but the mathematical resemblance of some of their laws will remain, and may still be made useful in exciting appropriate mathematical ideas.

"It is by the use of analogies of this kind that I have attempted to bring before the mind, in a convenient and manageable form, those mathematical ideas which are necessary to the study of the phenomena of electricity. The methods are generally those suggested by the processes of reasoning which are found in the researches of Faraday, and which, though they have been interpreted mathematically by Professor Thomson and others, are very generally supposed to be of an indefinite and unmathematical character, when compared with those employed by the professed mathematicians. By the method which I adopt, I hope to render it evident that I am not attempting to establish any physical theory of a science in which I have hardly made a single experiment, and that the limit of my design is to show how, by a strict application of the ideas and methods of Faraday, the connection of the very different orders of phenomena which he has discovered may be clearly placed before the mathematical mind. I shall therefore avoid as much as I can the introduction of anything which does not serve as a direct illustration of Faraday's methods, or of the mathematical deductions which may be made from them. In treating the simpler parts of the subject I shall use Faraday's mathematical methods as well as his ideas. When the complexity of the subject requires it, I shall use analytical notation, still confining myself to the development of ideas originated by the same philosopher.

"I have in the first place to explain and illustrate the idea of 'lines of force.'

"When a body is electrified in any manner, a small body

charged with positive electricity, and placed in any given position, will experience a force urging it in a certain direction. If the small body be now negatively electrified, it will be urged by an equal force in a direction exactly opposite.

"The same relations hold between a magnetic body and the north or south poles of a small magnet. If the north pole is urged in one direction, the south pole is urged in the opposite direction.

"In this way we might find a line passing through any point of space, such that it represents the direction of the force acting on a positively electrified particle, or on an elementary north pole, and the reverse direction of the force on a negatively electrified particle or an elementary south pole. Since at every point of space such a direction may be found, if we commence at any point and draw a line so that, as we go along it, its direction at any point shall always coincide with that of the resultant force at that point, this curve will indicate the direction of that force for every point through which it passes, and might be called on that account a *line of force*. We might in the same way draw other lines of force, till we had filled all space with curves indicating by their direction that of the force at any assigned point.

"We should thus obtain a geometrical model of the physical phenomena, which would tell us the *direction* of the force, but we should still require some method of indicating the *intensity* of the force at any point. If we consider these curves not as mere lines, but as fine tubes of variable section carrying an incompressible fluid, then, since the velocity of the fluid is inversely as the section of the tube, we may make the velocity vary according to any given law, by regulating the section of the tube, and in this way we might represent the intensity of the force as well as its direction by the motion of the fluid in these tubes. This method of representing the intensity of a force by the velocity of an imaginary fluid in a tube is applicable to any conceivable system of forces, but it is capable of great simplification in the case in which the forces are such as can be explained by the hypothesis of attractions varying inversely as the square of the distance, such as those observed in electrical and magnetic phenomena. In the case of a perfectly arbitrary system of forces, there will generally be interstices between the tubes; but in the case of electric and magnetic forces it is possible to arrange the tubes so as to leave no interstices. The tubes will then be mere surfaces, directing the motion of a fluid filling up the whole space. It has been usual to commence the investigation of the laws of these forces by at once assuming that the phenomena are due to attractive or repulsive forces acting between certain points. We may, however, obtain a different view of the subject, and one more suited to our more difficult inquiries, by adopting for the definition of the forces of which we treat, that they may be represented in magnitude and direction by the uniform motion of an incompressible fluid.

"I propose, then, first to describe a method by which the motion of such a fluid can be clearly conceived; secondly to trace the consequences of assuming certain conditions of motion, and to point out the application of the method to some of the less complicated phenomena of electricity, magnetism, and galvanism; and lastly, to show how by an extension of these methods, and the introduction of another idea due to Faraday, the laws of the attractions and inductive actions of magnets and currents may be clearly conceived, without making any assumptions as to the physical nature of electricity, or adding anything to that which has been already proved by experiment.

"By referring everything to the purely geometrical idea of the motion of an imaginary fluid, I hope to attain generality and precision, and to avoid the dangers arising from a premature theory professing to explain the cause of the phenomena. If the results of mere speculation which I have collected are found to be of any use to experimental philosophers, in arranging and interpreting their results, they will have served their purpose, and a mature theory, in which physical facts will be physically explained, will be formed by those who by interrogating Nature herself can obtain the only true solution of the questions which the mathematical theory suggests."

The idea was a bold one: for a youth of twenty-three to explain, by means of the motions of an incompressible fluid, some of the less complicated phenomena of electricity and magnetism, to show how the laws of the attractions of magnets and currents may be clearly conceived without making any assumption as to the physical nature of electricity, or adding anything to that which has already been proved by experiment.

It may be useful to review in a very few words the position of electrical theory* in 1855.

Coulomb's experiments had established the fundamental facts of electrostatic attraction and repulsion, and Coulomb himself, about 1785, had stated a theory based on these experiments which could "only be attacked by proving his experimental results to be inaccurate."†

Coulomb supposes the existence of two electric fluids, the theory developed previously by Franklin, but says—

"Je préviens pour mettre la théorie qui va suivre à l'abri de toute dispute systématique, que dans la supposition de deux fluides électriques, je n'ai autre intention que de présenter avec le moins d'éléments possible les résultats du calcul et de l'expérience, et non d'indiquer les véritables causes de l'électricité."

Cavendish was working in England about the same time as Coulomb, but he published very little,

* An historical account of the development of the science of electricity will be found in the article "Electricity" in the *Encyclopædia Britannica*, ninth edition, by Professor Chrystal.

† Thomson (Lord Kelvin), "Papers on Electrostatics and Magnetism," p. 15.

and the value and importance of his work was not recognised until the appearance in 1879 of the "Electrical Researches of Henry Cavendish," edited by Clerk Maxwell.

Early in the present century the application of mathematical analysis to electrical problems was begun by Laplace, who investigated the distribution of electricity on spheroids, and about 1811 Poisson's great work on the distribution of electricity on two spheres placed at any given distance apart was published. Meanwhile the properties of the electric current were being investigated. Galvani's discovery of the muscular contraction in a frog's leg, caused by the contact of dissimilar metals, was made in 1790. Volta invented the voltaic pile in 1800, and Oersted in 1820 discovered that an electric current produced magnetic force in its neighbourhood. On this Ampère laid the foundation of his theory of electro-dynamics, in which he showed how to calculate the forces between circuits carrying currents from an assumed law of force between each pair of elements of the circuits. His experiments proved that the consequences which follow from this law are consistent with all the observed facts. They do not prove that Ampère's law alone can explain the facts.

Maxwell, writing on this subject in the "Electricity and Magnetism," vol. ii., p. 162, says:—

"The experimental investigation by which Ampère established the laws of the mechanical action between electric currents is one of the most brilliant achievements in science.

"The whole, theory and experiment, seems as if it had leaped full grown and full armed from the brain of the

'Newton of Electricity.' It is perfect in form and unassailable in accuracy, and it is summed up in a formula from which all the phenomena may be deduced, and which must always remain the cardinal formula of electro-dynamics.

"The method of Ampère, however, though cast into an inductive form, does not allow us to trace the formation of the ideas which guided it. We can scarcely believe that Ampère really discovered the law of action by means of the experiments which he describes. We are led to suspect, what, indeed, he tells us himself, that he discovered the law by some process which he has not shown us, and that when he had afterwards built up a perfect demonstration, he removed all traces of the scaffolding by which he had built it."

The experimental evidence for Ampère's theory, so far, at least, as it was possible to obtain it from experiments on closed circuits, was rendered unimpeachable by W. Weber about 1846, while in the previous year Grassman and F. E. Neumann both published laws for the attraction between two elements of current which differ from that of Ampère, but lead to the same result for closed circuits. In a paper published in 1846 Weber announced his hypothesis connecting together electrostatic and electro-dynamic action. In this paper he supposed that the force between two particles of electricity depends on the motion of the particles as well as on their distance apart. A somewhat similar theory was proposed by Gauss and published after his death in his collected works. It has been shown, however, that Gauss' theory is inconsistent with the conservation of energy. Weber's theory avoids this inconsistency and leads, for closed circuits, to the same results as Ampère. It has been proved, however, by Von Helmholtz, that, under certain circumstances, according to it, a body would behave as though its mass were negative—it would move in a direction opposite to that of the force.*

Since 1846 many other theories have been proposed to explain Ampère's laws. Meanwhile, in 1821, Faraday observed that under certain circumstances a wire carrying a current could be kept in continuous rotation in a magnetic field by the action between the magnets and the current. In 1824 Arago observed the motion of a magnet caused by rotating a copper disc in its neighbourhood, while in 1831 Faraday began his experimental researches into electro-magnetic induction. About the same period Joseph Henry, of Washington, was making, independently of Faraday, experiments of fundamental importance on electro-magnetic induction, but sufficient attention was not called to his work until comparatively recent years.

In 1833 Lenz made some important researches, which led him to discover the connection between the direction of the induced currents and Ampère's laws, summed up in his rule that the direction of the induced current is always such as to oppose by its electro-magnetic action the motion which induces it.

In 1845 F. E. Neumann developed from this law the mathematical theory of electro-magnetic induction, and about the same time W. Weber showed how it might be deduced from his elementary law of electrical action.

The great name of Von Helmholtz first appears in connection with this subject in 1851, but of his writings we shall have more to say at a later stage.

* J. J. Thomson, B.A., Report, 1885, pp. 109, 113, Report on **Electrical Theories.**

Meanwhile, during the same period, various writers, Murphy, Plana, Charles, Sturm, and Gauss, extended Poisson's work on electrostatics, treating the questions which arose as problems in the distribution of an attracting fluid, attracting or repelling according to Newton's law, though here again the greatest advances were made by a self-taught Nottingham shoemaker, George Green by name, in his paper "On the Application of Mathematical Analysis to the Theories of Electricity and Magnetism," 1828.

Green's researches, Lord Kelvin writes, "have led to the elementary proposition which must constitute the legitimate foundation of every perfect mathematical structure that is to be made from the materials furnished by the experimental laws of Coulomb."

Green, it may be remarked, was the inventor of the term Potential. His essay, however, lay neglected from 1828, until Lord Kelvin called attention to it in 1845. Meanwhile, some of its most important results had been re-discovered by Gauss and Charles and Thomson himself.

Until about 1845, the experimental work on which these mathematical researches in electrostatics were based was that of Coulomb. An electrified body is supposed to have a charge of some imponderable fluid "electricity." Particles of electricity repel each other according to a certain law, and the fluid distributes itself in equilibrium over the surface of any charged conductor in accordance with this law. There are on this theory two opposite kinds of electric fluid, positive and negative, two charges of the same kind repel, two charges of opposite kinds attract; the repulsion or attraction is proportional to the product of the charges, and inversely proportional to the square of the distance between them.

The action between two charges is action at a distance taking place across the space which separates the two.

Faraday, in 1837, in the eleventh series of his "Experimental Researches," published his first paper on "Electrostatic Induction." He showed—as indeed Cavendish had proved long previously, though the result remained unpublished—that the force between two charged bodies will depend on the insulating medium which surrounds them, not merely on their shape and position. Induction, as he expresses it, takes place along curved lines, and is an action of contiguous particles; these curved lines he calls the "lines of force."

Discussing these researches in 1845, Lord Kelvin writes*:—

"Mr. Faraday's researches . . . were undertaken with a view to test an idea which he had long possessed that the forces of attraction and repulsion exercised by free electricity are not the resultants of actions exercised at a distance, but are propagated by means of molecular action among the contiguous particles of the insulating medium surrounding the electrified bodies, which he therefore calls the dielectric. By this idea he has been led to some very remarkable views upon induction, or, in fact, upon electrical action in general. As it is impossible that the phenomena observed by Faraday can be incompatible with the results of experiment which constitute Coulomb's theory, it is to be expected that the difference of his ideas from those of Coulomb must arise solely from a different method of stating and interpreting physically the

* Papers on "Electrostatics," etc., p. 26.

same laws; and further, it may, I think, be shown that either method of viewing this subject, when carried sufficiently far, may be made the foundation of a mathematical theory which would lead to the elementary principles of the other as consequences. This theory would, accordingly, be the expression of the ultimate law of the phenomena, independently of any physical hypothesis we might from other circumstances be led to adopt. That there are necessarily two distinct elementary ways of viewing the theory of electricity may be seen from the following considerations. . . ."

In the pages which follow, Lord Kelvin develops the consequences of an analogy between the conduction of heat and electrostatic action, which he had pointed out three years earlier (1842), in his paper on "The Uniform Motion of Heat in Homogeneous Solid Bodies," and discusses its connection with the mathematical theory of electricity.

The problem of distributing sources of heat in a given homogeneous conductor of heat, so as to produce a definite steady temperature at each point or the conductor is shewn to be *mathematically* identical with that of distributing electricity in equilibrium, so as to produce at each point an electrical potential having the same value as the temperature.

Thus the fundamental laws of the conduction of heat may be made the basis of the mathematical theory of electricity, but the physical idea which they suggest is that of the propagation of some effect by means of the mutual action of contiguous particles, rather than that of material particles attracting or repelling at a distance, which naturally follows from the statement of Coulomb's law.

Lord Kelvin continues:—

"All the views which Faraday has brought forward and illustrated, as demonstrated by experiment, lead to this method of establishing the mathematical theory, and, as far as the analysis is concerned, it would in most *general* propositions be more simple, if possible, than that of Coulomb. Of course the analysis of *particular* problems would be identical in the two methods. It is thus that Faraday arrives at a knowledge of some of the most important of the mathematical theorems which from their nature seemed destined never to be perceived except as mathematical truths."

Lord Kelvin's papers on "The Mathematical Theory of Electricity," published from 1848 to 1850, his "Propositions on the Theory of Attraction" (1842), his "Theory of Electrical Images" (1847), and his paper on "The Mathematical Theory of Magnetism" (1849), contain a statement of the most important results achieved in the mathematical sciences of Electrostatics and Magnetism up to the time of Maxwell's first paper.

The opening sentences of that paper have already been quoted. In the preface to the "Electricity and Magnetism" Maxwell writes thus:—

"Before I began the study of electricity I resolved to read no mathematics on the subject till I had first read through 'Experimental Researches on Electricity.' I was aware that there was supposed to be a difference between Faraday's way of conceiving phenomena and that of the mathematicians, so that neither he nor they were satisfied with each other's language. I had also the conviction that this discrepancy did not arise from either party being wrong. I was first convinced of this by Sir William Thomson, to whose advice and assistance, as well as to his published papers, I owe most of what I have learned on the subject.

"As I proceeded with the study of Faraday, I perceived that his method of conceiving the phenomena was also a

K

mathematical one, though not exhibited in the conventional form of mathematical symbols. I also found that these methods were capable of being expressed in the ordinary mathematical forms, and thus compared with those of the professed mathematicians.

"For instance, Faraday, in his mind's eye, saw lines of force traversing all space where the mathematicians saw centres of force attracting at a distance. Faraday saw a medium where they saw nothing but distance. Faraday sought the seat of the phenomena in real actions going on in the medium. They were satisfied that they had found it in a power of action at a distance impressed on the electric fluids."

Now, Maxwell saw an analogy between electrostatics and the steady motion of an incompressible fluid like water, and it is this analogy which he develops in the first part of his paper. The water flows along definite lines; a surface which consists wholly of such lines of flow will have the property that no water ever crosses it. In any stream of water we can imagine a number of such surfaces drawn, dividing it up into a series of tubes; each of these will be a tube of flow, each of these tubes remain always filled with water. Hence, the quantity of water which crosses per second any section of a tube of flow perpendicular to its length is always the same. Thus, from the form of the tube, we can obtain information as to the direction and strength of the flow, for where the tube is wide the flow will be proportionately small, and *vice versâ*.

Again, we can draw in the fluid a number of surfaces, over each of which the pressure is the same; these surfaces will cut the tubes of flow at right angles. Let us suppose they are drawn so that the difference of pressure between any two consecutive surfaces is unity, then the surfaces will be close together at points at which the pressure changes rapidly; where the variation of pressure is slow, the distance between two consecutive surfaces will be considerable.

If, then, in any case of motion, we can draw the pressure surfaces, and the tubes of flow, we can determine the motion of the fluid completely. Now, the same mathematical expressions which appear in the hydro-dynamical theory occur also in the theory of electricity, the meaning only of the symbols is changed. For velocity of fluid we have to write electrical force. For difference of fluid pressure we substitute work done, or difference of electrical potential or pressure.

The surfaces and tubes, drawn as the solution of any hydro-dynamical problem, give us also the solution of an electrical problem; the tubes of flow are Faraday's tubes of force, or tubes of induction, the surfaces of constant pressure are surfaces of equal electrical potential. Induction may take place in curved lines just as the tubes of flow may be bent and curved; the analogy between the two is a complete one.

But, as Maxwell shows, the analogy reaches further still. An electric current flowing along a wire had been recognised as having many properties similar to those of a current of liquid in a tube. When a steady current is passing through any solid conductor, there are formed in the conductor tubes of electrical flow and surfaces of constant pressure. These tubes and surfaces are the same as those formed by the flow of liquid through a solid whose boundary surface is the same as that of the conductor, provided the flow of liquid is properly proportioned to the flow of electricity.

These analogies refer to steady currents in which, therefore, the flow at any point of the conductor does not depend on the time. In Part II. of his paper Maxwell deals with Faraday's electro-tonic state. Faraday had found that when *changes* are produced in the magnetic phenomena surrounding a conductor, an electric current is set up in the conductor, which continues so long as the magnetic changes are in progress, but which ceases when the magnetic state becomes steady.

"Considerations of this kind led Professor Faraday to connect with his discovery of the induction of electric currents the conception of a state into which all bodies are thrown by the presence of magnets and currents. This state does not manifest itself by any known phenomena as long as it is undisturbed, but any change in this state is indicated by a current or tendency towards a current. To this state he gave the name of the 'Electro-tonic State,' and although he afterwards succeeded in explaining the phenomena which suggested it by means of less hypothetical conceptions, he has on several occasions hinted at the probability that some phenomena might be discovered which would render the electro-tonic state an object of legitimate induction. These speculations, into which Faraday had been led by the study of laws which he has well established, and which he abandoned only for want of experimental data for the direct proof of the unknown state, have not, I think, been made the subject of mathematical investigation. Perhaps it may be thought that the quantitative determinations of the various phenomena are not sufficiently rigorous to be made the basis of a mathematical theory. Faraday, however, has not contented himself with simply stating the numerical results of his experiments and leaving the law to be discovered by calculation. Where he has perceived a law he has at once stated it, in terms as unambiguous as those of pure mathematics, and if the mathematician, receiving this as a physical truth, deduces from it other laws capable of being tested by experiment, he has merely assisted the physicist in arranging his own ideas, which is confessedly a necessary step in scientific induction.

"In the following investigation, therefore, the laws established by Faraday will be assumed as true, and it will be shown that by following out his speculations other and more general laws can be deduced from them. If it should, then, appear that these laws, originally devised to include one set of phenomena, may be generalised so as to extend to phenomena of a different class, these mathematical connections may suggest to physicists the means of establishing physical connections, and thus mere speculation may be turned to account in experimental science."

Maxwell shows how to obtain a mathematical expression for Faraday's electro-tonic state. In his "Electricity and Magnetism," this electro-tonic state receives a new name. It is known as the Vector Potential,* and the paper under consideration contains,

* It is difficult to explain without analysis exactly what is measured by Maxwell's Vector Potential. Its rate of change at any point of space measures the electromotive force at that point, so far as it is due to variations of the electric current in neighbouring conductors; the magnetic induction depends on the first differential coefficients of the components of the electro-tonic state; the electric current is related to their second differential coefficients in the same manner as the density of attracting matter is related to the potential it produces. In language which is now frequently used in mathematical physics, the electromotive force at a point due to magnetic induction is proportioned to the rate of change of the Vector Potential, the magnetic induction depends on the "curl" of the Vector Potential, while the electric current is measured by the "concentration" of the Vector Potential. From a knowledge of the Vector Potential these other quantities can be obtained by processes of differentiation.

though in an incomplete form, his first statement of those equations of the electric field which are so indissolubly bound up with Maxwell's name.

The great advance in theory made in the paper is the distinct recognition of certain mathematical functions as representing Faraday's electrotonic-state, and their use in solving electro-magnetic problems.

The paper contains no new physical theory of electricity, but in a few years one appeared. In his later writings Maxwell adopted a more general view of the electro-magnetic field than that contained in his early papers on "Physical Lines of Force." It must, therefore, not be supposed that the somewhat gross conception of cog-wheels and pulleys, which we are about to describe, were anything more to their author than a model, which enabled him to realise how the changes, which occur when a current of electricity passes through a wire, might be represented by the motion of actual material particles.

The problem before him was to devise a physical theory of electricity, which would explain the forces exerted on electrified bodies by means of action between the contiguous parts of the medium in the space surrounding these bodies, rather than by direct action across the distance which separates them. A similar question, still unanswered, had arisen in the case of gravitation. Astronomers have determined the forces between attracting bodies: they do not know how those forces arise.

Maxwell's fondness for models has already been alluded to; it had led him to construct his top to illustrate the dynamics of a rigid body rotating about a fixed point, and his model of Saturn's rings (now in the Cavendish Laboratory) to illustrate the motion of the satellites in the rings. He had explained many of the gaseous laws by means of the impact of molecules, and now his fertile ingenuity was to imagine a mechanical model of the state of the electro-magnetic field near a system of conductors carrying currents.

Faraday, as we have seen, looked upon electro-static and magnetic induction as taking place along curved lines of force. He pictures these lines as ropes of molecules starting from a charged conductor, or a magnet, as the case may be, and acting on other bodies near. These ropes of molecules tend to shorten, and at the same time to swell outwards laterally. Thus the charged conductor tends to draw other bodies to itself, there is a tension along the lines of force, while at the same time each tube of molecules pushes its neighbours aside: a pressure at right angles to the lines of force is combined with this tension. Assuming for a moment this pressure and tension to exist, can we devise a mechanism to account for it? Maxwell himself has likened the lines of force to the fibres of a muscle. As the fibres contract, causing the limb to which they are attached to move, they swell outwards, and the muscle thickens.

Again, from another point of view, we might consider a line of force as consisting of a string of small cells of some flexible material each filled with fluid. If we then suppose this series of cells caused to rotate rapidly about the direction of the line of force, the cells will expand laterally and contract longitudinally; there will again be tension along the lines of force and pressure at right angles to them. It was this last idea, as we shall see shortly, of which Maxwell made use—

"I propose now" [he writes ("On Physical Lines of Force," *Phil. Mag.*, vol. xxi.)] "to examine magnetic phenomena from a mechanical point of view, and to determine what tensions in, or motions of, a medium are capable of producing the mechanical phenomena observed. If by the same hypothesis we can connect the phenomena of magnetic attraction with electro-magnetic phenomena, and with those of induced currents, we shall have found a theory which, if not true, can only be proved to be erroneous by experiments, which will greatly enlarge our knowledge of this part of physics."

Lord Kelvin had in 1847 given a mechanical representation of electric, magnetic and galvanic forces by means of the displacements of an elastic solid in a state of strain. The angular displacement at each point of the solid was taken as proportional to the magnetic force, and from this the relation between the various other electric quantities and the motion of the solid was developed. But Lord Kelvin did not attempt to explain the origin of the observed forces by the effects due to these strains, but merely made use of the mathematical analogy to assist the imagination in the study of both.

Maxwell considered magnetic action as existing in the form of pressure or tension, or more generally, of some stress in some medium. The existence of a medium capable of exerting force on material bodies and of withstanding considerable stress, both pressure and tension, is thus a fundamental hypothesis with him; this medium is to be capable of motion, and electro-magnetic forces arise from its motion and its stresses.

Now, Maxwell's fundamental supposition is that, in a magnetic field, there is a rotation of the molecules continually in progress about the lines of magnetic force. Consider now the case of a uniform magnetic field, whose direction is perpendicular to the paper; we are to look upon the lines of force as parallel strings of molecules, the axes of these strings being perpendicular to the paper. Each string is supposed to be rotating in the same direction about its axis, and the angular velocity of rotation is a measure of the magnetic force. In consequence of this rotation there will be differences of pressure in different directions in the medium; the pressure along the axes of the strings will be less than it would be if the medium were at rest, that in the directions at right angles to the axes will be greater, the medium will behave as though it were under tension along the axes of the molecules under pressure at right angles to them. Moreover, it can be shown that the pressure and the tension are both proportional to the square of the angular velocity—the square, that is, of the magnetic force—and this result is in accordance with the consequences of experiment.

More elaborate calculation shows that this statement is true generally. If we draw the lines of force in any magnetic field, and then suppose the molecules of the medium set in rotation about these lines of force as axes, with velocities which at each point are proportional to the magnetic force, the distribution of pressure throughout is that which we know actually to exist in the magnetic field.

According to this hypothesis, then, a permanent bar magnet has the power of setting the medium round it into continuous molecular rotation about the lines of force as axes. The molecules which are set in rotation we may consider as spherical, or nearly spherical, cells filled with a fluid, or an elastic solid substance, and surrounded by a kind of membrane, or sack, holding the contents together.

So far the model does not give any account of electrical actions which go on in the magnetic field.

The energy is wholly rotational, and the forces wholly magnetic.

Consider, however, any two contiguous strings of molecules. Let them cut the paper as shown in the two circles in Fig. 1 :—

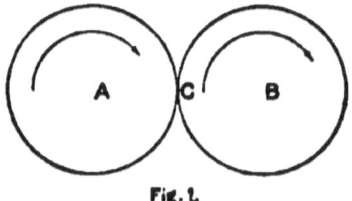 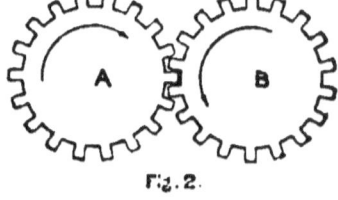

Fig. 1.   Fig. 2.

Then these cells are both rotating in the same direction, hence at A, where they touch, their points of contact will be moving in opposite directions, as shown by the arrow heads, and it is difficult to imagine how such motion can continue; it would require the surfaces of the cells to be perfectly smooth, and if this were so they would lose the power of transmitting action from one cell to the next.

The cells A and B may be compared to two cog-wheels placed close together, which we wish to turn in the same direction. If the cogs can interlock, as in Fig. 2, this is impossible: consecutive wheels in the train must move in opposite directions.

But in many machines the desired end is attained by inserting between the two wheels A and B a third idle wheel C, as shown in Fig. 3. This may be very

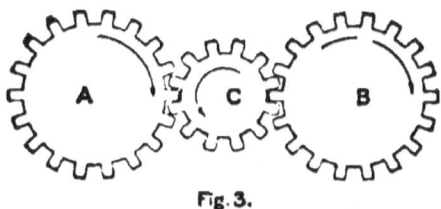

Fig. 3.

small, its only function is to transmit the motion of A to B in such a way that A and B may both turn in the same direction. It is not necessary that there should be cogs on the wheels: if the surfaces be perfectly rough, so that no slipping can take place, the same result follows without the cogs.

Guided by this analogy Maxwell extended his model by supposing each cell coated with a number of small particles which roll on its surface. These particles play the part of the idle wheels in the machine, and by their rolling merely enable the adjacent parts of two cells to move in opposite directions.

Consider now a number of such cells and their idle wheels lying in a plane, that of the paper, and suppose each cell is rotating with the same uniform angular velocity about an axis at right angles to that plane, each idle wheel will be acted on by two equal and opposite forces at the ends of the diameter in which it is touched by the adjacent cells; it will therefore be set in rotation, but there will be no force tending to drive it onwards; it does not matter whether the axis on which it rotates is free to move or fixed, in either case the idle wheel simply rotates. But suppose now the adjacent cells are not rotating at the same rate. In addition to its rotation the idle wheel will be urged onward with a velocity which depends on the difference between the rotations, and, if it can move freely, it will move on from between the two cells. Imagine now that the interstices between the cells are fitted with a string of idle wheels. So long as the adjacent cells move with different velocity there will be a continual stream of rolling particles or idle wheels between them. Maxwell in the paper considered these rolling particles to be particles of electricity. Their motion constitutes an electric current. In a uniform magnetic field there is no electric current; if the strength of the field varies, the idle wheels are set in motion and there may be a current.

These particles are very small compared with the magnetic vortices. The mass of all the particles is inappreciable compared with the mass of the vortices, and a great many vortices with their surrounding particles are contained in a molecule of the medium; the particles roll on the vortices without touching each other, so that so long as they remain within the same molecule there is no loss of energy by resistance. When, however, there is a current or general transference of particles in one direction they must pass from one molecule to another, and in doing so may experience resistance and generate heat.

Maxwell states that the conception of a particle, having its motion connected with that of a vortex by perfect rolling contact, may appear somewhat awkward. "I do not bring it forward," he writes, "as a mode of connection existing in Nature, or even as that which I would willingly assent to as an electrical hypothesis. It is, however, a mode of connection which is mechanically conceivable and easily investigated, and it serves to bring out the actual mechanical connections between the known electro-magnetic phenomena, so that I venture to say that anyone who understands the provisional and temporary character of this hypothesis will find himself rather helped than hindered by it in his search after the true interpretation of the phenomena."

The first part of the paper deals with the theory of magnetism; in the second part the hypothesis is applied to the phenomena of electric currents, and it is shown how the known laws of steady currents and of electro-magnetic induction can be deduced from it. In Part III., published January and February, 1862, the theory of molecular vortices is applied to statical electricity.

The distinction between a conductor and an insulator or dielectric is supposed to be that in the former the particles of electricity can pass with more or less freedom from molecule to molecule. In the latter such transference is impossible, the particles can only be displaced within the molecule with which they are connected; the cells or vortices of the medium are supposed to be elastic, and to resist by their elasticity the displacement of the particles within them. When electrical force acts on the medium this displacement of the particles within each molecule takes place until the stresses due to the elastic reaction of the vortices balance the electrical force; the medium behaves like an elastic body yielding to pressure until the pressure is balanced by the elastic stress. When the electric force is removed the cells or vortices recover their form, the electricity returns to its former position.

In a medium such as this waves of periodic displacement could be set up, and would travel with a velocity depending on its electric properties. The value for this velocity can be obtained from electrical observations, and Maxwell showed that this velocity, so found, was, within the limits of experimental error, the same as that of light. Moreover, the electrical oscillations take place, like those of light, in the front of the wave. Hence, he concludes, "the elasticity of the magnetic medium in air is the same as that of the luminiferous medium, if these two coexistent, coextensive, and equally elastic media are not rather one medium."

The paper thus contains the first germs of the electro-magnetic theory of light. Moreover, it is shown that the attraction between two small bodies charged with given quantities of electricity depends on the medium in which they are placed, while the specific inductive capacity is found to be proportional to the square of the refractive index.

The fourth and final part of the paper investigates the propagation of light in a magnetic field.

Faraday had shown that the direction of vibration

in a wave of polarised light travelling parallel to the lines of force in a magnetic field is rotated by its passage through the field. The numerical laws of this relation had been investigated by Verdet, and Maxwell showed how his hypothesis of molecular vortices led to laws which agree in the main with those found by Verdet.

He points out that the connection between magnetism and electricity has the same mathematical form as that between certain other pairs of phenomena, one of which has a *linear* and the other a *rotatory* character; and, further, that an analogy may be worked out assuming either the linear character for magnetism and the rotatory character for electricity, or the reverse. He alludes to Prof. Challis' theory, according to which magnetism is to consist in currents in a fluid whose directions correspond with the lines of magnetic force, while electric currents are supposed to be accompanied by, if not dependent upon, a rotatory motion of the fluid about the axis of the current; and to Von Helmholtz's theory of a somewhat similar character. He then gives his own reasons—agreeing with those of Sir W. Thomson (Lord Kelvin)—for supposing that there must be a real rotation going on in a magnetic field in order to account for the rotation of the plane of polarisation, and, accepting these reasons as valid, he develops the consequences of his theory with the results stated above.

His own verdict on the theory is given in the "Electricity and Magnetism" (vol. ii., § 831, first edition, p. 416):—

"A theory of molecular vortices, which I worked out at considerable length, was published in the *Phil. Mag.* for March, April, and May, 1861; Jan. and Feb., 1862.

"I think we have good evidence for the opinion that some phenomenon of rotation is going on in the magnetic field, that this rotation is performed by a great number of very small portions of matter, each rotating on its own axis, this axis being parallel to the direction of the magnetic force, and that the rotations of these different vortices are made to depend on one another by means of some kind of mechanism connecting them.

"The attempt which I then made to imagine a working model of this mechanism must be taken for no more than it really is, a demonstration that mechanism may be imagined capable of producing a connection mechanically equivalent to the actual connection of the parts of the electro-magnetic field. The problem of determining the mechanism required to establish a given species of connection between the motions of the parts of a system always admits of an infinite number of solutions. Of these, some may be more clumsy or more complex than others, but all must satisfy the conditions of mechanism in general.

"The following results of the theory, however, are of higher value:—

"(1) Magnetic force s the effect of the centrifugal force of the vortices.

"(2) Electro-magnetic induction of currents is the effect of the forces called into play when the velocity of the vortices is changing.

"(3) Electromotive force arises from the stress on the connecting mechanism.

"(4) Electric displacement arises from the elastic yielding of the connecting mechanism."

In studying this part of Maxwell's work, it must clearly be remembered that he did not look upon the ether as a series of cog-wheels with idle wheels between, or anything of the kind. He devised a mechanical model of such cogs and idle wheels, the properties of which would in some respects closely resemble those of the ether; from this model he deduced, among other things, the important fact that electric waves would travel outwards with the velocity of light. Other such models have been devised since his time to illustrate the same laws. Prof. Fitzgerald has actually constructed one of wheels connected together by elastic bands, which shows clearly the kind of processes which Maxwell supposed to go on in a dielectric when under electric force. Professor Lodge, in his book, "Modern Views of Electricity," has very fully developed a somewhat different arrangement of cog-wheels to attain the same result.

Maxwell's predictions as to the propagation of electric waves have in recent days received their full verification in the brilliant experiments of Hertz and his followers; it remains for us, before dealing with these, to trace their final development in his hands.

The papers we have been discussing were perhaps too material to receive the full attention they deserved; the ether is not a series of cogs, and electricity is something different from material idle wheels. In his paper on "The Dynamical Theory of the Electro-magnetic Field," *Phil. Trans.*, 1864, Maxwell treats the same questions in a more general manner. On a former occasion he says, "I have attempted to describe a particular kind of motion and a particular kind of strain so arranged as to account for the phenomena. In the present paper I avoid any hypothesis of this kind; and in using such words as electric momentum and electric elasticity in reference to the known phenomena of the induction of currents

L

and the polarisation of dielectrics, I wish merely to direct the mind of the reader to mechanical phenomena, which will assist him in understanding the electrical ones. All such phrases in the present paper are to be considered as illustrative and not as explanatory." He then continues:—

"In speaking of the energy of the field, however, I wish to be understood literally. All energy is the same as mechanical energy, whether it exists in the form of motion or in that of elasticity, or in any other form.

"The energy in electro-magnetic phenomena is mechanical energy. The only question is, Where does it reside?

"On the old theories it resides in the electrified bodies, conducting circuits, and magnets, in the form of an unknown quality called potential energy, or the power of producing certain effects at a distance. On our theory it resides in the electro-magnetic field, in the space surrounding the electrified and magnetic bodies, as well as in those bodies themselves, and is in two different forms, which may be described without hypothesis as magnetic polarisation and electric polarisation, or, according to a very probable hypothesis, as the motion and the strain of one and the same medium.

"The conclusions arrived at in the present paper are independent of this hypothesis, being deduced from experimental facts of three kinds:—

"(1) The induction of electric currents by the increase or diminution of neighbouring currents according to the changes in the lines of force passing through the circuit.

"(2) The distribution of magnetic intensity according to the variations of a magnetic potential.

"(3) The induction (or influence) of statical electricity through dielectrics.

"We may now proceed to demonstrate from these principles the existence and laws of the mechanical forces, which act upon electric currents, magnets, and electrified bodies placed in the electro-magnetic field."

In his introduction to the paper, he discusses in a general way the various explanations of electric phenomena which had been given, and points out that—

"It appears, therefore, that certain phenomena in electricity and magnetism lead to the same conclusion as those of optics, namely, that there is an ætherial medium pervading all bodies, and modified only in degree by their presence; that the parts of this medium are capable of being set in motion by electric currents and magnets; that this motion is communicated from one part of the medium to another by forces arising from the connection of those parts; that under the action of these forces there is a certain yielding depending on the elasticity of these connections; and that, therefore, energy in two different forms may exist in the medium, the one form being the actual energy of motion of its parts, and the other being the potential energy stored up in the connections in virtue of their elasticity.

"Thus, then, we are led to the conception of a complicated mechanism capable of a vast variety of motion, but at the same time so connected that the motion of one part depends, according to definite relations, on the motion of other parts, these motions being communicated by forces arising from the relative displacement of the connected parts, in virtue of their elasticity. Such a mechanism must be subject to the general laws of dynamics, and we ought to be able to work out all the consequences of its motion, provided we know the form of the relation between the motions of the parts."

These general laws of dynamics, applicable to the motion of any connected system, had been developed by Lagrange, and are expressed in his generalised equations of motion. It is one of Maxwell's chief claims to fame that he saw in the electric field a connected system to which Lagrange's equations could be applied, and that he was able to deduce the mechanical and electrical actions which take place by means of fundamental propositions of dynamics.

The methods of the paper now under discussion were developed further in the "Treatise on Electricity and Magnetism," published in 1873; in endeavouring to give some slight account of Maxwell's work, we shall describe it in the form it ultimately took.

The task which Maxwell set himself was a double one; he had first to express in symbols, in as general a form as possible, the fundamental laws of electromagnetism as deduced from experiments, chiefly the experiments of Faraday, and the relations between the various quantities involved; when this was done he had to show how these laws could be deduced from the general dynamical laws applicable to any system of moving bodies.

There are two classes of phenomena, electric and magnetic, which have been known from very early times, and which are connected together. When a piece of sealing-wax is rubbed it is found to attract other bodies, it is said to exert electric force throughout the space surrounding it; when two different metals are dipped in slightly acidulated water and connected by a wire, certain changes take place in the plates, the water, the wire, and the space round the wire, electric force is again exerted and a current of electricity is said to flow in the wire. Again, certain bodies, such as the lodestone, or pieces of iron and steel which have been treated in a certain manner, exhibit phenomena of action at a distance: they are said to exert magnetic force, and it is found that this magnetic force exists in the neighbourhood of an electric current and is connected with the current.

Again, when electric force is applied to a body, the

effects may be in part electrical, in part mechanical; the electrical state of the body is in general changed, while in addition, mechanical forces tending to move the body are set up. Experiment must teach us how the electrical state depends on the electric force, and what is the connection between this electric force and the magnetic forces which may, under certain circumstances, be observed. Now, in specifying the electric and magnetic conditions of the system, various other quantities, in addition to the electric force, will have to be introduced; the first step is to formulate the necessary quantities, and to determine the relations between them and the electric force.

Consider now a wire connecting the two poles of an electric battery—in its simplest form, a piece of zinc and a piece of copper in a vessel of dilute acid—electric force is produced at each point of the wire. Let us suppose this force known; an electric current depending on the material and the size of the wire flows along it, its value can be determined at each point of the wire in terms of the electric force by Ohm's law. If we take either this current or the electric force as known, we can determine by known laws the electric and magnetic conditions elsewhere. If we suppose the wire to be straight and very long, then, so long as the current is steady and we neglect the small effect due to the electrostatic charge on the wire, there is no electric force outside the wire. There is, however, magnetic force, and it is found that the lines of magnetic force are circles round the wire. It is found also that the work done in travelling once completely round the wire against the magnetic force is measured by the current flowing through the wire, and is obtained in the system of units usually adopted by multiplying the current by $4\pi$. This last result then gives us one of the necessary relations, that between the magnetic force due to a current and the strength of the current.

Again, consider a steady current flowing in a conductor of any form or shape, the total flow of current across any section of the conductor can be measured in various ways, and it is found that at any time this total flow is the same for each section of the conductor. In this respect the flow of a current resembles that of an incompressible fluid through a pipe; where the pipe is narrow the velocity of flow is greater than it is where the pipe is broad, but the total quantity crossing each section at any given instant is the same.

Consider now two conducting bodies, two spheres, or two flat plates placed near together but insulated. Let each conductor be connected to one of the poles of the battery by a conducting wire. Then, for a very short interval after the contact is made, it is found that there is a current in each wire which rapidly dies away to zero. In the neighbourhood of the balls there is electric force; the balls are said to be charged with electricity, and the lines of force are curved lines running from one ball to the other. It is found that the balls slightly attract each other, and the space between them is now in a different condition from what it was before the balls were charged. According to Maxwell, *Electric Displacement* has been produced in this space, and the electric displacement at each point is proportional to the electric force at that point.

Thus, (i) when electric force acts on a conductor, it produces a current, the current being by Ohm's law proportional to the force: (ii) when it acts on an insulator it produces electric displacement, and the displacement is proportional to the force; while (iii) there is magnetic force in the neighbourhood of the current, and the work done in carrying a magnetic pole round any complete circuit linked with the current is proportional to the current. The first two of these principles give us two sets of equations connecting together the electric force and the current in a conductor or the displacement in a dielectric respectively; the third connects the magnetic force and the current.

Now let us go back to the variable period when the current is flowing in the wires; and to make ideas precise, let the two conductors be two equal large flat plates placed with their faces parallel, and at some small distance apart. In this case, when the plates are charged, and the current has ceased, the electric displacement and the force are confined almost entirely to the space between the plates. During the variable period the total flow at any instant across each section of the wire is the same, but in the ordinary sense of the word there is no flow of electricity across the insulating medium between the plates. In this space, however, the electric displacement is continuously changing, rising from zero initially to its final steady value when the current ceases. It is a fundamental part of Maxwell's theory that this variation of electric displacement is equivalent in all respects to a current. The current at any point in a dielectric is measured by the rate of change of displacement at that point.

Moreover, it is also an essential point that if we consider any section of the dielectric between the two plates, the rate of change of the total displacement across this section is at each moment equal to the total flow of current across each section of the conducting wire.

Currents of electricity, therefore, including displacement currents, always flow in closed circuits, and obey the laws of an incompressible fluid in that the total flow across each section of the circuit —conducting or dielectric—is at any moment the same.

It should be clearly remembered that this fundamental hypothesis of Maxwell's theory is an assumption only to be justified by experiment. Von Helmholtz, in his paper on "The Equations of Motion of Electricity for Bodies at Rest," formed his equations in an entirely different manner from Maxwell, and arrived at results of a more general character, which do not require us to suppose that currents flow always in closed circuits, but permit of the condensation of electricity at points in the circuit where the conductors end and the non-conducting part of the circuit begins. We leave for the present the question which of the two theories, if either, represents the facts.

We have obtained above three fundamental relations—(i) that between electric force and electric current in a conductor; (ii) that between electric force and electric displacement in a dielectric; (iii) that between magnetic force and the current which gives rise to it. And we have seen that an electric current—*i.e.* in a dielectric the variation of the strength of an electric field of force—gives rise to magnetic force. Now, magnetic force acting on a medium produces "magnetic displacement," or magnetic induction, as it is called. In all media except iron, nickel, cobalt, and a few other substances, the magnetic induction is proportional to the magnetic force, and the ratio between the magnetic induction produced by a given force and the force is found to be very nearly the same for all such media. This ratio is known as the permeability, and is generally denoted by the symbol $\mu$.

A relation reciprocal to that given in (iii) above might be anticipated, and was, in fact, discovered by Faraday. Changes in a field of magnetic induction give rise to electric force, and hence to displacement currents in a dielectric or to conduction currents in a conductor. In considering the relation between these changes and the electric force, it is simplest at first not to deal with magnetic matter such as iron, nickel, or cobalt; and then we may say that (iv) the work which at any instant would be done in carrying a unit quantity of electricity round a closed circuit in a magnetic field against the electric forces due to the field is equal to the rate at which the total magnetic induction which threads the circuit is being decreased. This law, summing up Faraday's experiments on electro-magnetic induction, gives a fourth principle, leading to a fourth series of equations connecting together the electric and magnetic quantities involved.

The equations deduced from the above four principles, together with the condition implied in the continuity of an electric current, constitute Maxwell's equations of the electro-magnetic field.

If we are dealing only with a dielectric medium, the reciprocal relation between the third and fourth principle may be made more clear by the following statement :—

(A) The work done at any moment in carrying a unit quantity of magnetism round a closed circuit in a field in which electric displacement is varying, is equal to the rate of change of the total electric displacement through the circuit multiplied by $4\pi$.*

(B) The work done at any moment in carrying a unit quantity of electricity round a circuit in a field in which the magnetic induction is varying, is equal to the rate of change of the total magnetic induction through the circuit.

From these two principles, combined with the laws connecting electric force and displacement, magnetic force and induction, and with the condition of continuity, Maxwell obtained his equations of the field.

Faraday's experiments on electro-magnetic induction afford the proof of the truth of the fourth principle. It follows from those experiments that when the number of lines of magnetic induction

---

* The $4\pi$ is introduced because of the system of units usually employed to measure electrical quantities. If we adopted Mr. Oliver Heaviside's " rational units," it would disappear, as it does in (B).

which are linked with any closed circuit are made to vary, an induced electromotive force is brought into play round that circuit. This electromotive force is, according to Faraday's results, measured by the rate of decrease in the number of lines of magnetic induction which thread the circuit. Maxwell applies this principle to all circuits, whether conducting or not.

In obtaining equations to express in symbols the results of the fourth principle just enunciated, Maxwell introduces a new quantity, to which he gives the name of the "vector potential." This quantity appears in his analysis, and its physical meaning is not at first quite clear. Professor Poynting has, however, put Maxwell's principles in a slightly different form, which enables us to see definitely the meaning of the vector potential, and to deduce Maxwell's equations more readily from the fundamental statements.

We are dealing with a circuit with which lines of magnetic induction are linked, while the number of such lines linked with the circuit is varying. Now, let us suppose the variation to take place in consequence of the lines of induction moving outwards or inwards, as the case may be, so as to cut the circuit. Originally there are none linked with the circuit. As the magnetic field has grown to its present strength lines of magnetic induction have moved inwards. Each little element of the circuit has been cut by some, and the total number linked with the circuit can be found by adding together those cut by each element. Now, Professor Poynting's statement of Maxwell's fourth principle is that the electrical force in the direction of any element of the circuit is found by dividing by the length of the element the number of lines of magnetic induction which are cut in one second by it.

Moreover, the total number of lines of magnetic induction which have been cut by an element of unit length is defined as the component of the vector potential in the direction of the element; hence the electrical force in any direction is the rate of decrease of the component of the vector potential in that direction. We have thus a physical meaning for the vector potential, and shall find that in the dynamical theory this quantity is of great importance.

Professor Poynting has modified Maxwell's third principle in a similar manner; he looks upon the variation in the electric displacement as due to the motion of tubes of electric induction,* and the magnetic force along any circuit is equal to the number of tubes of electric induction cutting or cut by unit length of the circuit per second, multiplied by $4\pi$.

From the equations of the field, as found by Maxwell, it is possible to derive two sets of symmetrical equations. The one set connects the rate of change of the electric force with quantities depending on the magnetic force; the other set connects in a similar manner the rate of change of the magnetic force with quantities depending on the electric force.

* For an exact statement as to the relation between the directions of the lines of electric displacement and of the magnetic force, reference must be made to Professor Poynting's paper, *Phil. Trans.*, 1885, Part II., pp. 280, 281. The ideas are further developed in a series of articles in the *Electrician*, September, 1895. Reference should also be made to J. J. Thomson's "Recent Researches in Electricity and Magnetism."

Several writers in recent years adopt these equations as the fundamental relations of the field, establishing them by the argument that they lead to consequences which are found to be in accordance with experiment.

We have endeavoured to give some account of Maxwell's historical method, according to which the equations are deduced from the laws of electric currents and of electro-magnetic induction derived directly from experiment.

While the manner in which Maxwell obtained his equations is all his own, he was not alone in stating and discussing general equations of the electro-magnetic field. The next steps which we are about to consider are, however, in a special manner due to him. An electrical or magnetic system is the seat of energy; this energy is partly electrical, partly magnetic, and various expressions can be found for it. In Maxwell's theory it is a fundamental assumption that energy has position. "The electric and magnetic energies of any electro-magnetic system," says Professor Poynting, "reside, therefore, somewhere in the field." It follows from this that they are present wherever electric and magnetic force can be shown to exist. Maxwell showed that all the electric energy is accounted for by supposing that in the neighbourhood of a point at which the electric force is R there is an amount of energy per unit of volume equal to $KR^2/8\pi$, $K$ being the inductive capacity of the medium, while in the neighbourhood of a point at which the magnetic force is H, the magnetic energy per unit of volume is $\mu H^2/8\pi$, $\mu$ being the permeability. He supposes, then, that at each point of an electro-magnetic system energy is stored according to these laws. It follows, then, that the electro-magnetic field resembles a dynamical system in which energy is stored. Can we discover more of the mechanism by which the actions in the field are maintained? Now the motion of any point of a connected system depends on that of other points of the system; there are generally, in any machine, a certain number of points called driving-points, the motion of which controls the motion of all other parts of the machine; if the motion of the driving-points be known, that of any other point can be determined. Thus in a steam engine the motion of a point on the fly-wheel can be found if the motion of the piston and the connections between the piston and the wheel be known.

In order to determine the force which is acting on any part of the machine we must find its momentum, and then calculate the rate at which this momentum is being changed. This rate of change will give us the force. The method of calculation which it is necessary to employ was first given by Lagrange, and afterwards developed, with some modifications, by Hamilton. It is usually referred to as Hamilton's principle; when the equations in the original form are used they are known as Lagrange's equations.

Now Maxwell showed how these methods of calculation could be applied to the electro-magnetic field. The energy of a dynamical system is partly kinetic, partly potential. Maxwell supposes that the magnetic energy of the field is kinetic energy, the electric energy potential. When the kinetic energy of a system is known, the momentum of any part of the system can be calculated by recognised processes. Thus if we consider a circuit in an electro-magnetic field we can calculate the energy of the field, and hence obtain the momentum corresponding to this circuit. If we deal with a simple case in which the conducting circuits are fixed in position, and only the current in each circuit is allowed to vary, the rate of change of momentum corresponding to any circuit will give the force in that circuit. The momentum in question is electric momentum, and the force is electric force. Now we have already seen that the electric force at any point of a conducting circuit is given by the rate of change of the vector potential in the direction considered. Hence we are led to identify the vector potential with the electric momentum of our dynamical system; and, referring to the original definition of vector potential, we see that the electric momentum of a circuit is measured by the number of lines of magnetic induction which are interlinked with it.

Again, the kinetic energy of a dynamical system can be expressed in terms of the squares and products of the velocities of its several parts. It can also be expressed by multiplying the velocity of each driving-point by the momentum corresponding to that driving-point, and taking half the sum of the products. Suppose, now, we are dealing with a system consisting of a number of wire circuits in which currents are running, and let us suppose that we may represent the current in each wire as the velocity of a driving-point in our dynamical system. We can also express in terms of these currents the electric momentum of each wire circuit; let this be done, and let half the sum of the products of the corresponding velocities and momenta be formed.

In maintaining the currents in the wires energy is needed to supply the heat which is produced in each wire; but in starting the currents it is found that more energy is needed than is requisite for the supply of this heat. This excess of energy can be calculated, and when the calculation is made it is found that the excess is equal to half the sum of the products of the currents and corresponding momenta. Moreover, if this sum be expressed in terms of the magnetic force, it is found to be equal to $\mu H^2/8\pi$, which is the magnetic energy of the field. Now, when a dynamical system is set in motion against known forces, more energy is supplied than is needed to do the work against the forces; this excess of energy measures the kinetic energy acquired by the system.

Hence, Maxwell was justified in taking the magnetic energy of the field as the kinetic energy of the mechanical system, and if the strengths of the currents in the wires be taken to represent the velocities of the driving-points, this energy is measured in terms of the electrical velocities and momenta in exactly the same way as the energy of a mechanical system is measured in terms of the velocities and momenta of its driving-points.

The mechanical system in which, according to Maxwell, the energy is stored is the ether. A state of motion or of strain is set up in the ether of the field. The electric forces which drive the currents, and also the mechanical forces acting on the conductors carrying the currents, are due to this state of motion, or it may be of strain, in the ether. It must not be supposed that the term electric displacement in Maxwell's mind meant an actual bodily displacement of the particles of the ether; it is in some way connected with such a material displacement. In his view, without motion of the ether particles there would be no electric action, but he does not identify electric displacement and the displacement of an ether particle.

His mechanical theory, however, does account for the electro-magnetic forces between conductors carrying currents. The energy of the system depends on the relative positions of the currents which form part of it. Now, any conservative mechanical system tends to set itself in such a position that its potential energy is least, its kinetic energy greatest. The circuits of the system, then, will tend to set themselves so that the electro-kinetic energy of the system may be as large as possible; forces will be needed to hold them in any position in which this condition is not satisfied.

We have another proof of the correctness of the value found for the energy of the field in that the forces calculated from this value agree with those which are determined by direct experiment.

Again, the forces applied at the various driving-points are transmitted to other points by the connections of the machine; the connections are thrown into a state of strain; stress exists throughout their substance. When we see the piston-rod and the shaft of an engine connected by the crank and the connecting-rod, we recognise that the work done on the piston is transmitted thus to the shaft. So, too, in the electro-magnetic field, the ether forms the connection between the various circuits in the field; the forces with which those circuits act on each other are transmitted from one circuit to another by the stresses set up in the ether.

To take another instance, consider the electrostatic attraction between two charged bodies. Let us suppose the bodies charged by connecting each to the opposite pole of a battery; a current flows from the battery setting up electric displacement in the space between the bodies, and throwing the ether into a state of strain. As the strain increases the current gets less; the reaction resulting from the strain tends to stop it, until at last this reaction is so great that the current is stopped. When this is the case the wires to the battery may be removed, provided this is done without destroying the insulation of the bodies; the state of strain will remain and shows itself in the attraction between the balls.

Looking at the problem in this manner, we are face to face with two great questions—the one, What is the state of strain in the ether which will enable it to produce the observed electro-static attractions and repulsions between charged bodies? and the other, What is the mechanical structure of the ether which would give rise to such a state of strain as will account for the observed forces? Maxwell gives one answer to the first question; it is not the only answer which could be given, but it does account for the facts. He failed to answer the second. He says ("Electricity and Magnetism," vol. i. p. 132):—

"It must be carefully borne in mind that we have made only one step in the theory of the action of the medium. We have supposed it to be in a state of stress, but have not in any way accounted for this stress, or explained how it is maintained. ... I have not been able to make the next step, namely, to account by mechanical considerations for these stresses in the dielectric."

Faraday had pointed out that the inductive action between two bodies takes place along the lines of force, which tend to shorten along their length and to spread outwards in other directions. Maxwell compares them to the fibres of a muscle, which contracts and at the same time thickens when exerting force. In the electric field there is, on Maxwell's theory, a tension along the lines of electric force and a pressure at right angles to those lines. Maxwell proved that a tension $K R^2/8\pi$ along the lines of force, combined with an equal pressure in perpendicular directions, would maintain the equilibrium of the field, and would give rise to the observed attractions or repulsions between electrified bodies. Other distributions of stress might be found which would lead to the same result. The one just stated will always be connected with Maxwell's name. It will be noticed that the tension along the lines of force and the pressure at right angles to them are each numerically equal to the potential energy stored per unit of volume in the field. The value of each of the three quantities is $K R^2/8\pi$.

In the same way, in a magnetic field, there is a state of stress, and on Maxwell's theory this, too, consists of a tension along the lines of force and an equal pressure at right angles to them, the values of the tension and the pressure being each equal to that of the magnetic energy per unit of volume, or $\mu H^2/8\pi$.

In a case in which both electric and magnetic force exists, these two states of stress are superposed. The total energy per unit of volume is $K R^2/8\pi + \mu H^2/8\pi$; the total stress is made up of tensions $K R^2/8\pi$ and $\mu H^2/8\pi$ along the lines of electric and magnetic force respectively, and equal pressures at right angles to these lines.

We see, then, from Maxwell's theory, that electric force produced at any given point in space is transmitted from that point by the action of the ether. The question suggests itself, Does the transmission take time, and if so, does it proceed with a definite velocity depending on the nature of the medium through which the change is proceeding?

According to the molecular-vortex theory, we have seen that waves of electric force are transmitted with a definite velocity. The more general theory developed in the "Electricity and Magnetism" leads to the same result. Electric force produced at any point travels outwards from that point with a velocity given by $1/\sqrt{K\mu}$. At a distant point the force is zero, until the disturbance reaches it. If the disturbance last only for a limited interval, its effects will at any future time be confined to the space within a spherical shell of constant thickness depending on the interval; the radii of this shell increase with uniform speed $1/\sqrt{K\mu}$.

If the initial disturbance be periodic, periodic waves of electric force will travel out from the centre, just as waves of sound travel out from a bell, or waves of light from a candle flame. A wire carrying an alternating current may be such a source of periodic disturbance, and from the wire waves travel outwards into space.

Now, it is known that in a sound wave the displacements of the air particles take place in the direction in which the wave is travelling; they lie at right angles to the wave front, and are spoken of as longitudinal. In light waves, on the other hand, the displacements are, as Fresnel proved, in the wave front, at right angles, that is, to the direction of propagation; they are transverse.

Theory shows that in general both these waves may exist in an elastic solid body, and that they travel with different velocities. Of which nature are the waves of electric displacement in a dielectric? It can be shewn to follow as a necessary consequence of Maxwell's views as to the closed character of all electric currents, that waves of electric displacement are transverse. Electric vibrations, like those of light, are in the wave front and at right angles to the direction of propagation; they depend on the rigidity or quasi-rigidity of the medium through which they travel, not on its resistance to compression.

Again, an electric current, whether due to variation of displacement in a dielectric or to conduction in a conductor, is accompanied by magnetic force. A wave of periodic electric displacement, then, will be also a wave of periodic magnetic force travelling at the same rate; and Maxwell shewed that the direction of this magnetic force also lies in the wave front, and is always at right angles to the electric displacement. In the ordinary theory of light the wave of linear displacement is accompanied by a wave of periodic angular twist about a direction lying in the wave front and perpendicular to the linear displacement.

In many respects, then, waves of electric displacement resemble waves of light, and, indeed, as we proceed we shall find closer connections still. Hence comes Maxwell's electro-magnetic theory of light.

It is only in dielectric media that electric force is propagated by wave motion. In conductors, although the third and fourth of Maxwell's principles given on page 185 still are true, the relation between the electric force and the electric current differs from that which holds in a dielectric. Hence the equations satisfied by the force are different. The laws of its propagation resemble those of the conduction of heat rather than those of the transmission of light.

Again, light travels with different velocities in different transparent media. The velocity of electric waves, as has been stated, is equal to $1/\sqrt{\mu K}$; but in making this statement it is assumed that the simple laws which hold where there is no gross matter—or, rather, where air is the only dielectric with which we are concerned—hold also in solid or liquid dielectrics. In a solid or a liquid, as in vacuo, the waves are propagated by the ether. We assume, as a first step towards a complete theory, that so far as the electric waves are concerned the sole effect produced by the matter shews itself in a change of inductive capacity or of permeability. It is not likely that such a supposition should be the whole truth, and we may, therefore, expect results deduced from it to be only approximation to the true result.

Now, electro-magnetic experiments show that, excluding magnetic substances, the permeability of all bodies is very nearly the same, and differs very slightly from that of air. The inductive capacity, however, of different bodies is different, and hence the velocity with which electro-magnetic waves travel differs in different bodies.

But the refraction of waves of light depends on the fact that light travels with different velocities in different media; hence we should expect to have waves of electric displacement reflected and refracted when they pass from one dielectric, such as air, to another, such as glass or gutta-percha; moreover, for light the refractive index of a medium such as glass is the ratio of the velocity in air to the velocity in the glass.

Thus the electrical refractive index of glass is the ratio of the velocity of electric waves in air to their velocity in glass.

Now let $K_0$ be the inductive capacity of air, $K_1$ that of glass, taking the permeability of air and glass to be the same, we have the result that—

$$\text{Electrical refractive index} = \sqrt{K_1 / K_0}.$$

But the ratio of the inductive capacity of glass to that of air is known as the specific inductive capacity of glass.

Hence, the specific inductive capacity of any medium is equal to the square of the electrical refractive index of that medium.

Since Maxwell's time the mathematical laws of the reflexion and refraction of electric waves have been investigated by various writers, and it has been shown that they agree exactly with those enunciated by Fresnel for light.

Hitherto we have been discussing the propagation of electric waves in an isotropic medium, one which has identical properties in all directions about a point. Let us now consider how these laws are modified if the dielectric be crystalline in structure.

Maxwell assumes that the crystalline character of the dielectric can be sufficiently represented by supposing the inductive capacity to be different in different directions; experiments have since shown that this is true for crystals such as Iceland Spar and Aragonite; he assumes also, and this, too, is justified by experiment, that the magnetic permeability does not depend on the direction. It follows from these assumptions that a crystal will produce double refraction and polarisation of electric waves which fall upon it, and, further, that the laws of double refraction will be those given by Fresnel for light waves in a doubly refracting medium. There will be two waves in the crystal. The disturbance in each of these will be plane polarised; their velocity and the position of their plane of polarisation can be found from the direction in which they are travelling by Fresnel's construction exactly.

Maxwell's theory, then, would appear to indicate

some close connection between electric waves and those of light. Faraday's experiments on the rotation of the plane of polarisation by magnetic force shew one phenomenon in which the two are connected, and Maxwell endeavoured to apply his theory to explain this. Here, however, it became necessary to introduce an additional hypothesis—there must be some connection between the motion of the ether to which magnetic force is due and that which constitutes light. It is impossible to give a mechanical account of the rotation of the plane of polarisation without some assumption as to the relation between these two kinds of motion. Maxwell, therefore, supposes the linear displacements of a point in the ether to be those which give rise to light, while the components of the magnetic force are connected with these in the same way as the components of a vortex in a liquid in vortex motion are connected with the displacements of the liquid. He further assumes the existence of a term of special form in the expression for the kinetic energy, and from these assumptions he deduces the laws of the propagation of polarised light in a magnetic field. These laws agree in the main with the results of Verdet's experiments.

# CHAPTER X.

### DEVELOPMENT OF MAXWELL'S THEORY.

WE have endeavoured in the preceding pages to give some account of Maxwell's contributions to electrical theory and the physics of the ether. We must now consider very briefly what evidence there is to support these views. At Maxwell's death such evidence, though strong, was indirect. His supporters were limited to some few English-speaking pupils, young and enthusiastic, who were convinced, it may be, in no small measure, by the affection and reverence with which they regarded their master. Abroad his views had made very little way.

In the last words of his book he writes, speaking of various distinguished workers—

"There appears to be in the minds of these eminent men some prejudice, or *à priori* objection, against the hypothesis of a medium in which the phenomena of radiation of light and heat, and the electric actions at a distance, take place. It is true that, at one time, those who speculated as to the causes of physical phenomena were in the habit of accounting for each kind of action at a distance by means of a special æthereal fluid, whose function and property it was to produce these actions. They filled all space three and four times over with æthers of different kinds, the properties of which were invented merely to 'save appearances,' so that more rational enquirers were willing rather to accept not only Newton's definite law of attraction at a distance, but even the dogma of Cotes,[*] that action at a distance is one of the primary properties of matter, and that no explanation can be more intelligible than this fact. Hence the undulatory theory of light

[*] Preface to Newton's "Principia," 2nd edition.

has met with much opposition, directed not against its failure to explain the phenomena, but against its assumption of the existence of a medium in which light is propagated.

"We have seen that the mathematical expression for electro-dynamic action led, in the mind of Gauss, to the conviction that a theory of the propagation of electric action in time would be found to be the very key-stone of electro-dynamics. Now we are unable to conceive of propagation in time, except either as the flight of a material substance through space, or as the propagation of a condition of motion, or stress, in a medium already existing in space.

"In the theory of Neumann, the mathematical conception called potential, which we are unable to conceive as a material substance, is supposed to be projected from one particle to another in a manner which is quite independent of a medium, and which, as Neumann has himself pointed out, is extremely different from that of the propagation of light.

"In the theories of Riemann and Betti it would appear that the action is supposed to be propagated in a manner somewhat more similar to that of light.

"But in all of these theories the question naturally occurs :— If something is transmitted from one particle to another at a distance, what is its condition after it has left one particle and before it has reached the other? If this something is the potential energy of the two particles, as in Neumann's theory, how are we to conceive this energy as existing in a point of space, coinciding neither with the one particle nor with the other? In fact, whenever energy is transmitted from one body to another in time, there must be a medium or substance in which the energy exists after it leaves one body and before it reaches the other, for energy, as Torricelli\* remarked, 'is a quintessence of so subtle a nature that it cannot be contained in any vessel except the inmost substance of material things.' Hence all these theories lead to a conception of a medium in which the propagation takes place, and if we admit this medium as an hypothesis, I think it ought to occupy a prominent place in our investigations, and that we ought to

\* "Lezioni Accademiche" (Firenze, 1715), p. 25.

endeavour to construct a mental representation of all the details of its action, and this has been my constant aim in this treatise."

Let us see, then, what were the experimental grounds in Maxwell's day for accepting as true his views on electrical action, and how since then, by the genius of Heinrich Hertz and the labours of his followers, those grounds have been rendered so sure that nearly the whole progress of electrical science during the last twenty years has consisted in the development of ideas which are to be found in the "Treatise on Electricity and Magnetism."

The purely electrical consequences of Maxwell's theory were of course in accord with all known electrical observations. The equations of the field accounted for the electro-magnetic forces observed in various experiments, and from them the laws of electromagnetic induction could be correctly deduced; but there was nothing very special in this. Similar equations had been obtained from the theory of action at a distance by various writers; in fact, Helmholtz's theory, based on the most general form of expression for the force between two elements of current consistent with certain experiments of Ampère's, was more general in its character than Maxwell's. The destructive features of Maxwell's theory were:

(1) The assumption that all currents flow in closed circuits.

(2) The idea of energy residing throughout the electro-magnetic field in consequence of the strains and stresses set up in the electro-magnetic medium by the actions to which it was subject.

(3) The identification of this electro-magnetic medium with the luminiferous ether, and the consequent view that light is an electro-magnetic phenomena.

(4) The view that electro-magnetic forces arise entirely from strains and stresses set up in the ether; the electro-static charge of an insulated conductor being one of the forms in which the ether strain is manifested to us.

(5) A dielectric under the action of electric force is said to become polarised, and, according to Maxwell (vol. i. p. 133), all electrification is the residual effect of the polarisation of the dielectric.

Now it must, I think, be admitted that in Maxwell's day there was direct proof of very few of these propositions. No one has even yet so measured the displacement currents in a dielectric as to show that the total flow across every section of a circuit is at any given moment the same, though there are other experiments of an indirect character which have now completely justified Maxwell's hypothesis. Experiments by Schiller and Von Helmholtz prove it is true that some action in the dielectric must be taken into consideration in any satisfactory theory; they therefore upset various theories based on direct action at a distance, "but they tell us nothing as to whether any special form of the dielectric theory, such as Maxwell's or Helmholtz's, is true or not." (J. J. Thomson, "Report on Electrical Theories," B.A. Report, 1885, p. 149.)

When Maxwell died there had been little if any experimental evidence as to the stresses set up in a body by electric force. Fontana, Govi, and Duter had all observed that changes take place in the volume of the dielectric of a condenser when it is charged. Quincke had taken up the work, and the first of his classic papers on this subject was published in 1880, the year following Maxwell's death. Maxwell himself was fond of shewing an experiment in which a charged insulated sphere was brought near to the surface of paraffin; the stress on the surface causes a heaping up of the paraffin under the sphere.

Kerr had shewn in 1875 that many substances become doubly refracting under electric stress; his complete determination of the laws of this action was published at a later date.

As to direct measurements on electric waves, there were none; the value of the velocity with which, if Maxwell's theory were true, they must travel had been determined from electrical observations of quite a different character. Weber and Kohlrausch had measured the value of K for air, for which $\mu$ is unity, and from their observations it follows that the value of the wave velocity for electro-magnetic waves is about $31 \times 10^9$ centimetres per second. The velocity of light was known, from the experiments of Fizeau and Foucault, to have about this value, and it was the near coincidence of these two values which led Maxwell to write in 1864:—

"The agreement of the results seems to show that light and magnetism are affections of the same substance, and that light is an electro-magnetic disturbance propagated through the field according to electro-magnetic laws."

By the time the first edition of the "Electricity and Magnetism" was published, Maxwell and Thomson (Lord Kelvin) had both made determinations of K, and had shewn that for air at least the resulting value for the velocity of electro-magnetic waves was very nearly that of light.

For other substances at that date the observations were fewer still. Gibson and Barclay had determined the specific inductive capacity of paraffin, and found that its square root was 1·405, while its refractive index for long waves is 1·422. Maxwell himself thought that if a similar agreement could be shewn to hold for a number of substances, we should be warranted in concluding that "the square root of K, though it may not be the complete expression for the index of refraction, is at least the most important term in it."

Between this time and Maxwell's death enough had been done to more than justify this statement. It was clear from the observations of Boltzmann, Silow, Hopkinson, and others that there were many substances for which the square root of the specific inductive capacity was very nearly indeed equal to the refractive index, and good reason had been given why in some cases there should be a considerable difference between the two.

Hopkinson found that in the case of glass the differences were very large, and they have since been found to be considerable for most solids examined, with the exception of paraffin and sulphur. For petroleum oil, benzine, toluene, carbon-bisulphide, and some other liquids the agreement between Maxwell's theory and experiment is close. For the fatty oils, such as castor oil, olive oil, sperm oil, neatsfoot oil, and also for ether, the differences are considerable.

It seems probable that the reason for this difference lies in the fact that, in the light waves, we are dealing with the wave velocity of a disturbance of an extremely short period. Now, we know that the substances mentioned shew optical dispersion, and we have at present no completely satisfactory theory from which we can calculate, from experiments on very short waves, what the velocity for very long waves will be. In most cases Cauchy's formula has been used to obtain the numbers given. The value of $K$, however, as found by experiment, corresponds to these infinitely long waves, and to quote Professor J. J. Thomson's words, "the marvel is not that there should not be substances for which the relation $K = \mu^2$ does not hold, but that there should be any for which it does." *

It has been shewn, moreover, both by Professor J. J. Thomson himself and by Blondlot, that when the value of $K$ is measured under very rapidly varying electrifications, changing at the rate of about 25,000,000 to the second, the value of the inductive capacity for glass is reduced from about 6·8 or 7 to about 2·7; the square root of this is 1·6, which does not differ much from its refractive index. The values of the inductive capacity of paraffin and sulphur, which it will be remembered agree fairly with Maxwell's theory, were found to be not greatly different in the steady and in the rapidly varying field.

On the other hand, some experiments of Arons

* In his sentence $\mu$ stands for the refractive index.

and Rubens in rapidly varying fields lead to values which do not differ greatly from those given by other methods. The theory, however, of these experiments seems open to criticism.

To attempt anything like a complete account of modern verifications of Maxwell's views and modern developments of his theory is a task beyond our limits, but an account of Maxwell written in 1895 would be incomplete without a reference to the work of Heinrich Hertz.

Maxwell told us what the properties of electromagnetic waves in air must be. Hertz* in 1887 enabled us to measure those properties, and the measurements have verified completely Maxwell's views.

The method of producing electrical oscillations in a conductor had long been known. Thomson and Von Helmholtz had both pointed it out. Schiller had examined such oscillations in 1874, and had determined the inductive capacity of glass by their means, using oscillations whose period varied from ·000056 to ·00012 of a second.

These oscillations were produced by discharging a condenser through a coil of wire having self-induction. If the electrical resistance of the coil be not too great, the charge oscillates backwards and forwards between the plates of the condenser until its energy is dissipated in the heat produced in the wire, and in the electro-magnetic radiations which leave it.

The period of these oscillations under proper conditions is given by the formula $T = 2\pi\sqrt{CL}$ where

* Hertz's papers have been translated into English by D. E. Jones, and are published under the title of *Electric Waves*.

L, the coefficient of self induction, and C the capacity of the condenser. These quantities can be calculated, and hence the time of an oscillation is known. From such an arrangement waves radiate out into space. If we could measure by any method the length of such a wave we could determine its velocity by dividing the wave length by the period. But it is clear that since the velocity is comparable with that of light the wave length will be enormous, unless the period is very short. Thus, a wave, travelling with the velocity of light, whose period was ·0001 second, such as the waves Schiller worked with, would have a length of ·0001 × 30,000,000,000 or 3,000,000 centimetres, and would be quite unmeasurable. Before measurements on electric waves could be made it was necessary (1) to produce waves of sufficiently rapid period, (2) to devise means to detect them. This is what Hertz did.

The wave length of the electrical oscillations can be reduced by reducing either the electrical capacity of the system, or the coefficient of self-induction of the wire. Hertz adopted both these expedients. His vibrator, in some of his more important experiments, consisted of two square brass plates 40 cm. in the side. To each of these is attached a piece of copper wire about 30 cm. in length, and each wire ends in a small highly-polished brass ball. The plates are placed so that the wires lie in the same straight line, the brass balls being separated by a very small air gap. The two plates are then charged, the one positively the other negatively, until the insulation resistance of the air gap breaks down and a discharge passes across. Under these conditions the discharge is oscillatory. It does not consist of a single spark, but of a series of sparks, which pass and repass in opposite directions, until the energy of the original charge is radiated into space or dissipated as heat; the plates are then recharged and the process repeated. In Hertz's experiments the oscillator was charged by being connected to the secondary terminals of an induction coil.

In 1883 Professor Fitzgerald had called attention to this method of producing electric waves in air, and had given two metres as the minimum wave length which might be attained. In 1870 Herr von Bezold had actually made observations on the propagation and reflection of electrical oscillations, but his work, published as a preliminary communication, had attracted little notice. Hertz was the first to undertake in 1887 in a systematic manner the investigation of the electric waves in air which proceed from such an oscillator with a view to testing various theories of electro-magnetic action.

It remained, however, necessary to devise an apparatus for detecting the waves. When the waves are incident on a conductor, electric surgings are set up in the conductor, and may, under proper conditions, be observed as tiny sparks. Hertz used as his detector a loop of wire, the ends of which terminated in two small brass balls. The wire was bent so that the balls were very close together, and the sparks could be seen passing across the tiny air gap which separated them. Such a wire will have a definite period of its own for oscillations of electricity with which it may be charged, and if the frequency of the electric waves which fall on it agrees with that of the waves which it can itself emit, the oscillations which are set up in the wire will be stronger than under other conditions, the sparks seen will be more brilliant.* Hertz's resonator was a circle of wire thirty-five centimetres in radius, the period for such a resonator would, he calculated, be the same as that of his vibrator.

There is, however, very considerable difficulty in determining the period of an electric oscillator from its dimensions, and the value obtained from calculation for that of Hertz's radiator is not very trustworthy. The complete period is, however, comparable with two one hundredth millionths of a second; in his original papers, Hertz, through an error, gave a value greater than this.

With these arrangements Hertz was able to detect the presence of electrical radiation at considerable distances from the radiator; he was also able to measure its wave length. In the case of sound waves the existence of nodes and loops formed under proper conditions is well known. When waves are directly reflected from a flat surface, interference takes place between the incident and reflected waves, stationary vibrations are set up, and nodes and loops—places, that is, of minimum and of maximum motion respectively— are formed. The position of these nodes and loops can be determined by the aid of suitable apparatus, and it can be shewn that the distance between two consecutive nodes is half the wave length.

---

* Some of the consequences of this electrical resonance have been very strikingly shown by Professor Oliver Lodge. *See Nature, February 20th, 1890.*

Similarly when electrical vibrations fall on a reflector, a large flat surface of metal, for example, stationary vibrations due to the interference between the incident and reflected waves are produced, and these give rise to electrical nodes and loops. The position of such nodes and loops can be found by the use of Hertz's apparatus, or in other ways, and hence the length of the electrical waves can be found. The existence of the nodes and loops shews that the electric effects are propagated by wave motion. The length of the waves is found to be definite, since the nodes and loops recur at equal intervals apart.

If it be assumed that the frequency is known, the velocity of wave propagation can be determined. Hertz found from his experiments that in air the waves travelled with the velocity of light. It appears, however, that there were two errors in the calculation which happened to correct each other, so that neither the value of the frequency given in Hertz's paper nor the wave length observed is correct.

By modifying the apparatus it was possible to measure the wave length of the waves transmitted along a copper wire, and hence, again assuming the period of oscillation, to calculate the velocity of wave propagation along the wire. Hertz made the experiment, and found from his first observations that the waves were propagated along the wire with a finite velocity, but that the velocity differed from that in air. The half-wave length in the wire was only about 2·8 metres; that in air was about 4·5 metres.

Now, this experiment afforded a crucial test between the theories of Maxwell and Von Helmholtz.

According to the former, the waves do not travel in the wire at all; they travel through the air alongside the wire, and the wave length observed by Hertz ought to have been the same as in air. According to Von Helmholtz, the two velocities observed by Hertz should have been different, as, indeed, they were, and the experiment appeared to prove that Maxwell's theory was insufficient and that a more general one, such as that of Von Helmholtz, was necessary. But other experiments have not led to the same result. Hertz himself, using more rapid oscillations in some later measurements, found that the wave length of the electric waves from a given oscillator was the same whether they were transmitted through free space or conducted along a wire.* Lecher and J. J. Thomson have arrived at the same result; but the most complete experiments on this point are those of Sarasin and De la Rive.

It may be taken, then, as established that Maxwell's theory is sufficient, and that the greater generality of Von Helmholtz is unnecessary.

In a later paper Hertz showed that electric waves could be reflected and refracted, polarised and analysed, just like light waves. In his introduction to his " Collected Papers " he writes (p. 19) :—

"Casting now a glance backwards, we see that by the experiments above sketched the propagation in time of a

* Hertz's original results were no doubt affected by waves reflected from the walls and floor of the room in which he worked. An iron stove also, which was near his apparatus, may have had a disturbing influence; but for all this, it is to his genius and his brilliant achievements that the complete establishment of Maxwell's theory is due.

supposed action at a distance is for the first time proved. This fact forms the philosophic result of the experiments, and indeed, in a certain sense, the most important result. The proof includes a recognition of the fact that the electric forces can disentangle themselves from material bodies, and can continue to subsist as conditions or changes in the state of space. The details of the experiments further prove that the particular manner in which the electric force is propagated exhibits the closest analogy * with the propagation of light; indeed, that it corresponds almost completely to it. The hypothesis that light is an electrical phenomenon is thus made highly probable. To give a strict proof of this hypothesis would logically require experiments upon light itself.

"What we here indicate as having been accomplished by the experiments is accomplished independently of the correctness of particular theories. Nevertheless, there is an obvious connection between the experiments and the theory in connection with which they were really undertaken. Since the year 1861 science has been in possession of a theory which Maxwell constructed upon Faraday's views, and which we therefore call the Faraday-Maxwell theory. This theory affirms the possibility of the class of phenomena here discovered just as positively as the remaining electrical theories are compelled to deny it. From the outset Maxwell's theory excelled all others in elegance and in the abundance of the relations between the various phenomena which it included.

"The probability of this theory, and therefore the number of its adherents, increased from year to year. But as long as Maxwell's theory depended solely upon the probability of its results, and not on the certainty of its hypotheses, it could not completely displace the theories which were opposed to it.

"The fundamental hypotheses of Maxwell's theory contradicted the usual views, and did not rest upon the evidence of decisive experiments. In this connection we can best characterise the object and the result of our experiments by

* The analogy does not consist only in the agreement between the more or less accurately measured velocities. The approximately equal velocity is only one element among many others.

saying: The object of these experiments was to test the fundamental hypotheses of the Faraday-Maxwell theory, and the result of the experiments is to confirm the fundamental hypotheses of the theory."

Since Maxwell's death volumes have been written on electrical questions, which have all been inspired by his work. The standpoint from which electrical theory is regarded has been entirely changed. The greatest masters of mathematical physics have found, in the development of Maxwell's views, a task that called for all their powers, and the harvest of new truths which has been garnered has proved most rich. But while this is so, the question is still often asked, What is Maxwell's theory? Hertz himself concludes the introduction just referred to with his most interesting answer to this question. Prof. Boltzmann has made the theory the subject of an important course of lectures. Poincaré, in the introduction to his "Lectures on Maxwell's Theories and the Electromagnetic Theory of Light," expresses the difficulty, which many feel, in understanding what the theory is. "The first time," he says, "that a French reader opens Maxwell's book a feeling of uneasiness, often even of distrust, is mingled with his admiration. It is only after prolonged study, and at the cost of many efforts, that this feeling is dissipated. Some great minds retain it always." And again he writes: "A French *savant*, one of those who have most completely fathomed Maxwell's meaning, said to me once, 'I understand everything in the book except what is meant by a body charged with electricity.'"

In considering this question, Poincaré's own

remark—"Maxwell does not give a mechanical explanation of electricity and magnetism, he is only concerned to show that such an explanation is possible"—is most important.

We cannot find in the "Electricity" an answer to the question—What is an electric charge? Maxwell did not pretend to know, and the attempt to give too great definiteness to his views on this point is apt to lead to a misconception of what those views were.

On the old theories of action at a distance and of electric and magnetic fluids attracting according to known laws, it was easy to be mechanical. It was only necessary to investigate the manner in which such fluids could distribute themselves so as to be in equilibrium, and to calculate the forces arising from the distribution. The problem of assigning such a mechanical structure to the ether as will permit of its exerting the action which occurs in an electromagnetic field is a harder one to solve, and till it is solved the question—What is an electric charge?—must remain unanswered. Still, in order to grasp Maxwell's theory this knowledge is not necessary.

The properties of ether in dielectrics and in conductors must be quite different. In a dielectric the ether has the power of storing energy by some change in its configuration or its structure; in a conductor this power is absent, owing probably to the action of the matter of which the conductor is composed.

When we are said to charge an insulated conductor we really act on the ether in the neighbourhood of the body so as to store it with energy; if there be another conductor in the field we cannot store energy in the ether it contains. As, then, we pass from the outside of this conductor to its interior there is a sudden change in some mechanical quantity connected with the ether, and this change shows itself as a force of attraction between the two conductors. Maxwell called the change in structure, or in property, which occurs when a dielectric is thus stored with electro-static energy, *Electric Displacement;* if we denote it by D, then the electric force R is equal to $4\pi D/K$, and hence the energy in a unit of volume is $2\pi D^2/K$, where K is a quantity depending on the insulator.

Now, D, the electric displacement, is a quantity which has direction as well as magnitude. Its value, therefore, at any point can be represented by a straight line in the usual way; inside a conductor it is zero. The total change in D, which takes place all over the surface of a conductor as we enter it from the outside measures, according to Maxwell, the total charge on the conductor. At points at which the lines representing D enter the conductor the charge is negative; at points at which they leave it the charge is positive; along the lines of the displacement there exists throughout the ether a tension measured by $2\pi D^2/K$; at right angles to these lines there is a pressure of the same amount.

In addition to the above the components of the displacement D must satisfy certain relations which can only be expressed in mathematical form, the physical meaning of which it is difficult to state in non-mathematical language.

When these relations are so expressed the problem of finding the value of the displacement at all points of space becomes determinate, and the forces acting on the conductors can be obtained. Moreover, the total change of displacement on entering or leaving a conductor can be calculated, and this gives the quantity which is known as the total electrical charge on the conductor. The forces obtained by the above method are exactly the same as those which would exist if we supposed each conductor to be charged in the ordinary sense with the quantities just found, and to attract or repel according to the ordinary laws.

If, then, we define electric displacement as that change which takes place in a dielectric when it becomes the seat of electrostatic energy, and if, further, we suppose that the change, whatever it be mechanically, satisfies certain well-known laws, and that in consequence certain pressures and tensions exist in the dielectric, electrostatic problems can be solved without reference to a charge of electricity residing on the conductors.

Something such as this, it appears to me, is Maxwell's theory of electricity as applied to electrostatics. It is not necessary, in order to understand it, to know what change in the ether constitutes electric displacement, or what is an electric charge, though, of course, such knowledge would render our views more definite, and would make the theory a mechanical one.

When we turn to magnetism and electro-magnetism, Maxwell's theory develops itself naturally. Experiment proves that magnetic induction is connected with the rate of change of electric displacement, according to the laws already given. If, then, we knew the nature of the change to which the name "electric displacement" has been given, the nature of magnetic induction would be known. The difficulties in the way of any mechanical explanation are, it is true, very great; assuming, however, that some mechanical conception of "electric displacement" is possible, Maxwell's theory gives a consistent account of the other phenomena of electro-magnetism.

Again, we have, it is true, an electro-magnetic theory of light, but we do not know the nature of the change in the ether which affects our eyes with the sensation of light. Is it the same as electric displacement, or as magnetic induction, or since, when electric displacement is varying, magnetic induction always accompanies it, is the sensation of light due to the combined effect of the two?

These questions remain unanswered. It may be that light is neither electric displacement nor magnetic induction, but some quite different periodic change of structure of the ether, which travels through the ether at the same rate as these quantities, and obeys many of the same laws.

In this respect there is a material difference between the ordinary theory of light and the electromagnetic theory. The former is a mechanical theory; it starts from the assumption that the periodic change which constitutes light is the ordinary linear displacement of a medium—the ether—having certain mechanical properties, and from those properties it deduces the laws of optics with more or less success.

Lord Kelvin, in his labile ether, has devised a medium which could exist and which has the necessary mechanical properties. The periodic linear displacements of the labile ether would obey the laws of light, and from the fundamental hypotheses of the theory, a mechanical explanation, reasonably satisfactory in its main features, can be given of most purely optical phenomena. The relations between light and electricity, or light and magnetism, are not, however, touched by this theory: indeed, they cannot be touched without making some assumption as to what electric displacement is.

In recent years various suggestions have been made as to the nature of the change which constitutes electric displacement. One theory, due to Von Helmholtz, supposes that the electro-kinetic momentum, or vector potential of Maxwell, is actually the momentum of the moving ether; according to another, suggested, it would appear originally in a crude form by Challis, and developed within the last few months in very satisfactory detail by Larmor, the velocity of the ether is magnetic force; others have been devised, but we are still waiting for a second Newton to give us a theory of the ether which shall include the facts of electricity and magnetism, luminous radiation, and it may be gravitation.*

Meanwhile we believe that Maxwell has taken the first steps towards this discovery, and has pointed out the lines along which the future discoverer must direct his search, and hence we claim for him a foremost place among the leaders of this century of science.

---

* For a very suggestive account of some possible theories, reference should be made to the presidential address of Professor W. M. Hicks to Section A of the British Association at Ipswich in 1895.

# INDEX.

Aberdeen, Maxwell elected Professor at, 45; formation of University of, 51
Adams, W. G., succeeds Maxwell as Professor at King's College, London, 58
Adams Prize, The, 48; gained by Maxwell, 50
Ampère, 155, 204
Ampère's Law, 155, 156
*Annals of Philosophy*, Thomson's, 112, 113
"Apostles," club so called, 30, 89
Arago, 157
Arragonite, 200
Atom, article by Maxwell in *Encyclopædia Britannica*, 108
Avogadros' Law, 117, 124

Bakerian Lecture, delivered by Maxwell, 58
Berkeley on the Theory of Vision, 38
Bernouilli, D., 113
Blackburne, Professor, 16
Blore, Rev. E. W., 67
Boehm, Bust of Maxwell by, 90
Boltzmann, Dr., 135, 137, 138, 144, 216
Boltzmann-Maxwell Theory, The, 140, 145
Boscovitch on Atoms, 108, 109
Boyle's Law, 114, 117, 124
Brewster, Sir David, on Colour Sensation, 99
British Association, Maxwell and, 42, 54; Lecture before, 80-82; Lines on President's address, 83, 84
Butler, Dr. H. M., extract from sermon on Maxwell, 32-35
Bryan, G. H., 141, 143

Cambridge, Maxwell at, 28-46; Mathematical Tripos at, 60; Foundation of Professorship of Experimental Physics at, 66
*Cambridge and Dublin Mathematical Journal*, Papers by Maxwell in, 30
Campbell, Professor L., 9, 10, 12, 14, 22, 52, 57, 79
Cauchy's Formula, 208
Cavendish, Henry, 73, 74; Works of, edited by Maxwell, 87, 154, 155
Cavendish Laboratory, built and presented to University of Cambridge, 73, 74
Cay, Miss Frances, 11
Cayley Portrait Fund, lines to Committee, 86
Challis, Professor, 49
Charles' Law, 124
Chemical Society, Maxwell's lecture before, 80-82

Clausius, on kinetic theory of gases, 119, 129, 130, 137
Clerks of Penicuik, The, 9, 10
Colour Perception, 94
Colour Sensation, Young on, 97, 98; Sir D. Brewster on, 99
Colours, paper by Maxwell, on, 40, 41; Helmholtz on, 99
Conductors and Insulators, Distinction between, 173
Cookson, Dr., 61
Corsock, Maxwell buried at, 90
Cotes, 202
Coulomb, 154
Curves, investigated by Maxwell, 19

Daniell's cells, 77
Democritus, 108
Demonstrator of Physics, W. Garnett appointed, 75
Description of Oval Curves, first paper by Maxwell 19
Devonshire, Duke of, Cavendish Laboratory built by, 73, 74; Letter of Thanks from University of Cambridge, 74
Dewar, Miss K. M., her marriage to Maxwell, 51
Dickinson, Lowes; Portrait of Maxwell by, 90
Diffusion of gases, 128
Discs for colour experiments, 99-101
Droop, H. R., 57
Dynamical Theory of the Electromagnetic Field, Maxwell on, 57, 177
Dynamical Theory of Gases, Maxwell on, 58, 134

Edinburgh Academy, Maxwell's school-life at, 13-18
Edinburgh, Royal Society of, Maxwell at meetings of, 18
Edinburgh, University of, Maxwell at, 22
Elastic Spheres, 144
Electric Displacement, 218, 219, 220
Electrical Theories, 94, 154, 155
Electricity and Magnetism, Maxwell's book on, 59, 77, 79, 147, 155, 156, 176, 180-201; papers by Lord Kelvin on, 161-2; Application of Mathematical Analysis to, paper by G. Green, 158
Electricity, Modern Views of, by Professor Lodge, 177
Electro-kinetic Momentum, 221
Electro-magnetic Field, Dynamical Theory of, Maxwell on, 57, 177
Electro-magnetic Induction, 157
Electro-magnetic Theory of Light, 174

# INDEX. 223

Electro-tonic State, 164
Electrostatic Induction, Faraday on, 159
*Encyclopædia Britannica*, articles by Maxwell in, 80, 108, 146
Ether, labile, 220
Experimental Physics, foundation of Professorship at Cambridge, 66; Election of Maxwell, 68

Faraday on electrical science, 157; on electrostatic induction, 159
Faraday's Lines of Force, paper by Maxwell on, 44, 45, 148-153
Fawcett, W. M., architect of Cavendish Laboratory, 73
Fitzgerald, Professor, 177, 211
Forbes, Professor J. D., 18, 44, 54; friendship with Maxwell, 19; paper on Theory of Glaciers, 19; resigns Professorship at Edinburgh, 54

Galvani, 155
Garnett, W., appointed Demonstrator of Physics at Cambridge, 75; Life of Maxwell by, 94
Gases, Molecular theory of, 57, 108; Waterston on general theory of, 118; Clausius on, 119; diffusion of, 128
Gauss' Theory, 156
Gay Lussac's Law, 117
General Theory of Gases, Waterston on, 118; Clausius on, 119
Glenlair, home of Maxwell, 11, 23; laboratory at, 24; Maxwell's life at, 58, 59; "Electricity and Magnetism" written at, 79
Gordon, J. E. H., 77, 78
Green, G., of Nottingham, paper on electricity and magnetism, 158; inventor of term "Potential," 158

Hamilton, Sir W. R., 22
Hamilton's Principle, 190
Heat, Text-book on, by Maxwell, 79
Helmholtz, 99, 156, 157, 175, 221
Henry, J., of Washington, on electromagnetic induction, 157
Herapath on molecules, 112-116
Hertz, Heinrich, 204, 209-213
Hicks, W. M., 221
Hockin, C., 56
Holman, Professor, 133

Iceland Spar, 200
Insulators and Conductors, Distinction between, 173

Jenkin, Fleeming, 55, 56

Kelland, Professor, 22
Kelvin, Lord, 16, 142, 158, 159, 160, 168; on the Uniform Motion of Heat, 160;

papers on Electricity and Magnetism, 161, 162
Kinetic energy, 124, 129, 136, 139, 191
King's College, London, Maxwell elected Professor at, 54
Kohlrausch, 206
Kundt, 132

Labile Ether, 220
Laboratory at Glenlair, 24
Lagrange, 179
Lagrange's Equations, 179, 190
Laplace, 155
Larmor, J., 141, 142
Lecher, 214
Lenz, 157
Litchfield, R. B., 46
Light, Electro-magnetic Theory of, 174; Waves of, 198, 199
Lodge, Professor, book on Modern Views of Electricity, 177
Lucretius, 108
Luminous Radiation, 221

Mathematical Tripos at Cambridge, subjects, 60; Maxwell an examiner for, 60, 80; experimental work in, 76
Matter and Motion, Maxwell on, 79
Maxwell, James Clerk, parentage and birthplace, 10, 11; childhood and school-days, 12-18; his mother's death, 13; first lessons in geometry, 17; attends meetings of Royal Society of Edinburgh, 18; his first published paper, 19; friendship with Professor Forbes, 19; his polariscope, 20; enters the University of Edinburgh, 22; papers on Rolling Curves and Elastic Solids, 23; vacations at Glenlair, 23; laboratory at Glenlair, 24; undergraduate life at Cambridge, 28-36; elected scholar of Trinity, 29; illness at Lowestoft, 29; his friends at Cambridge, 30; Tripos and degree, 35-37; early researches, 38-44; paper on Colours, 40, 41; elected Fellow of Trinity, 43; Lecturer at Trinity, 43; Professor at Aberdeen, 45; his father's death, 45; gains the Adams Prize, 50; marriage, 51; powers as teacher and lecturer, 52, 53; Professor at King's College, London, 54; gains the Rumford Medal, 55; delivers Bakerian lecture, 58; resigns Professorship at King's College, London, 58; life at Glenlair, 58, 59; visit to Italy, 59; Examiner for Mathematical Tripos, 60, 80; elected Professor of Experimental Physics at Cambridge, 68; Introductory Lecture, 68-72; Examiner for Natural Sciences Tripos, 79; articles

in *Encyclopædia Britannica*, 60, 118, 146; papers in *Nature*, 80; lectures before British Association and Chemical Society, 80–82; humorous poems, 83–87; delivers Rede Lecture on the Telephone, 89; last illness and death, 89, 90; buried at Corsock, 90; bust and portrait, 90; religious views, 91, 92
Maxwell, John Clerk, 10, 11
Meyer, O. E., 133
Mill's Logic, 38
Molecular Evolution, Lines on, 85
—— Physics, 94
—— Constitution of Bodies, Maxwell on, 146
—— Theory of Gases, 57, 108
Molecules, 109, 110; Hempath on, 112–116; lecture by Maxwell on, 146
Motion of Saturn's Rings, subject for Adams Prize, 49
Munro, J. C., 40, 56, 68, 82

Natural Sciences Tripos, Maxwell Examiner for, 79
*Nature*, papers by Maxwell in, 80
Neumann, F. E., 156, 157
Newton's Lunar Theory and Astronomy, 50
—— Principia, 202
Nicol, Wm., inventor of the polarising prism, 20
Niven, W. D., 27, 46, 51, 52, 60, 78, 87, 88, 93

Obermeyer, 134
Ohm's Law, 77
Ophthalmoscope devised by Maxwell, 83
Oval Curves, Description of, Maxwell's first paper, 19

Parkinson, Dr., 49
*Philosophical Magazine*, 56, 99, 115, 120, 133, 142
*Philosophical Transactions*, 56, 89, 132, 145
Physical Lines of Force, Maxwell on, 56, 158
Physics, Instruction in, at Cambridge, 61; Report of Syndicate on, 62–64; Demonstrator appointed, 75
Poincaré, 216
Poisson, 44; on distribution of electricity, 155
Polariscope, made by Maxwell, 20
"Potential," term invented by G. Green, 158; the Vector, 165, 221
Poynting, Professor, 187–189
Puluj, 134

Quincke, 206

Radiation, Luminous, 221
Rarefied Gases, Stresses in, paper by Maxwell, 135, 145
Rayleigh, Lord, 67, 77
Rede Lecture on the Telephone, delivered by Maxwell, 89
Report on Electrical Theories, J. J. Thomson, 204
—— of Syndicate as to instruction in Physics at Cambridge, 62–64
Robertson, C. H., 28
Rolling Curves, Maxwell on, 23
Royal Society, The, Maxwell and, 55; Transactions of, 89
Rumford Medal gained by Maxwell, 55, 106

Sabine, Major-General, Vice-President of Royal Society, 106
Smith's Prizes, 36
Standards of Electrical Resistance, Committee on, 55
Stewart, Balfour, 56, 125
Stresses in Rarefied Gases, Maxwell on, 135, 155

Tait, Professor P. G., 21, 26, 94
Tayler, Rev. C. B., 29
Telephone, Rede Lecture by Maxwell on, 89
Theory of Glaciers, Prof. Forbes on, 19
Thomson, J. J., 157, 208; Report on Electrical Theories, 205
Thomson's *Annals of Philosophy*, 112, 113

Uniform Motion of Heat in Homogeneous Solid Bodies, paper by Lord Kelvin, 160, 161
University Commission, 47, 48, 62
Urr, Vale of, 11

Vector Potential, The, 165, 221
Viscosity of Gases, Experiments on, 58, 125, 132
Volta, Inventor of voltaic pile, 155

Waterston, J. J., on molecular theory of gases, 114, 115; on general theory of gases, 118
Waves of Light, 198, 199
Weber, W., 156, 206
Wedderburn, Mrs., 14
Wheatstone's Bridge, 77
Williams, J., Archdeacon of Cardigan, 16
Willis, Professor, 44
Wilson, E., lines in memory of, 86, 87

Young, T., on colour sensation, 97, 98

---

PRINTED BY CASSELL & COMPANY, LIMITED, LA BELLE SAUVAGE, LONDON, E.C.

*Selections from Cassell & Company's Publications.*

## Illustrated, Fine-Art, and other Volumes.

**Abbeys and Churches of England and Wales, The:** Descriptive, Historical, Pictorial. Series II. 21s.
**Adventure, The World of.** Fully Illustrated. In Three Vols. 9s. each.
**Africa and its Explorers, The Story of.** By Dr. ROBERT BROWN, F.L.S. Illustrated. Complete in 4 Vols., 7s. 6d. each.
**Animals, Popular History of.** By HENRY SCHERREN, F.Z.S. With 12 Coloured Plates and other Illustrations. 7s. 6d.
**Arabian Nights Entertainments, Cassell's Pictorial.** 10s. 6d.
**Architectural Drawing.** By R. PHENÉ SPIERS. Illustrated. 10s. 6d.
**Art, The Magazine of.** Yearly Vol. With 14 Photogravures or Etchings, a Series of Full-page Plates, and about 400 Illustrations. 21s.
**Artistic Anatomy.** By Prof. M. DUVAL. *Cheap Edition.* 3s. 6d.
**Astronomy, The Dawn of.** A Study of the Temple Worship and Mythology of the Ancient Egyptians. By Prof. J. NORMAN LOCKYER, C.B., F.R.S., &c. Illustrated. 21s.
**Atlas, The Universal.** A New and Complete General Atlas of the World, with 117 Pages of Maps, in Colours, and a Complete Index to about 125,000 Names. List of Maps, Prices and all Particulars on Application.
**Bashkirtseff, Marie, The Journal of.** *Cheap Edition.* 7s. 6d.
**Bashkirtseff, Marie, The Letters of.** 7s. 6d.
**Battles of the Nineteenth Century.** An Entirely New and Original Work. Illustrated. Vol. I., 9s.
**Beetles, Butterflies, Moths, and Other Insects.** By A. W. KAPPEL, F.E.S., and W. EGMONT KIRBY. With 12 Coloured Plates. 3s. 6d.
**"Belle Sauvage" Library, The.** Cloth, 2s. each. A list of the Volumes post free on application.
**Biographical Dictionary, Cassell's New.** *Cheap Edition*, 3s. 6d.
**Birds' Nests, Eggs, and Egg-Collecting.** By R. KEARTON. Illustrated with 16 Coloured Plates. 5s.
**Birds' Nests, British: How, Where, and When to Find and Identify Them.** By R. KEARTON. With an Introduction by Dr. BOWDLER SHARPE and upwards of 120 Illustrations of Nests, Eggs, Young, etc., from Photographs by C. KEARTON. 21s.
**Breech-Loader, The, and How to Use It.** By W. W. GREENER. Illustrated. New and enlarged edition. 2s. 6d.
**Britain's Roll of Glory;** or, the Victoria Cross, its Heroes, and their Valour. By D. H. PARRY. Illustrated. 7s. 6d.
**British Ballads.** With Several Hundred Original Illustrations. Complete in Two Vols., cloth, 15s. Half morocco, *price on application.*
**British Battles on Land and Sea.** By JAMES GRANT. With about 600 Illustrations. Four Vols., 4to, £1 16s.; *Library Edition*, £2.
**Butterflies and Moths, European.** With 61 Coloured Plates. 35s.
**Canaries and Cage-Birds, The Illustrated Book of.** With 56 Facsimile Coloured Plates, 35s. Half-morocco, £2 5s.
**Captain Horn, The Adventures of.** By FRANK STOCKTON. 6s.
**Capture of the "Estrella," The.** A Tale of the Slave Trade. By COMMANDER CLAUDE HARDING, R.N. 5s.
**Cassell's Family Magazine.** Yearly Vol. Illustrated. 7s. 6d.
**Cathedrals, Abbeys, and Churches of England and Wales.** Descriptive, Historical, Pictorial. *Popular Edition.* Two Vols. 25s.
**Cats and Kittens.** By HENRIETTE RONNER. With Portrait and 13 Full-page Photogravure Plates and numerous Illustrations. £2 2s.
**Chums.** The Illustrated Paper for Boys. Yearly Volume, 8s.
**Cities of the World.** Four Vols. Illustrated. 7s. 6d. each.
**Civil Service, Guide to Employment in the.** Entirely New Edition Paper, 1s. Cloth, 1s. 6d.
**Clinical Manuals for Practitioners and Students of Medicine.** A List of Volumes forwarded post free on application to the Publishers.

5 G. 8.95

*Selections from Cassell & Company's Publications.*

---

**Colour.** By Prof. A. H. CHURCH. With Coloured Plates. 3s. 6d.
**Commons and Forests, English.** By the Rt. Hon. G. SHAW-LEFEVRE, M.P. With Maps. 10s 6d.
**Cook, The Thorough Good.** By GEORGE AUGUSTUS SALA. 21s.
**Cookery, A Year's.** By PHYLLIS BROWNE. 3s. 6d.
**Cookery Book, Cassell's New Universal.** By LIZZIE HERITAGE. With 12 Coloured Plates and other Illustrations. Strongly bound in Half-leather. 1,344 pages. 6s.
**Cookery, Cassell's Shilling.** 110*th Thousand.* 1s.
**Cookery, Vegetarian.** By A. G. PAYNE. 1s. 6d.
**Cooking by Gas, The Art of.** By MARIE J. SUGG. Illustrated. 2s.
**Cottage Gardening, Poultry, Bees, Allotments, Etc.** Edited by W. ROBINSON. Illustrated. Half-yearly Volumes, 2s. 6d. each.
**Count Cavour and Madame de Circourt.** Some Unpublished Correspondence. Translated by A. J. BUTLER. Cloth gilt, 10s. 6d.
**Countries of the World, The.** By ROBERT BROWN, M.A., Ph.D., &c. *Cheap Edition.* Profusely Illustrated. Vol. I., 6s.
**Cyclopædia, Cassell's Concise.** Brought down to the latest date. With about 600 Illustrations. *Cheap Edition.* 7s. 6d.
**Cyclopædia, Cassell's Miniature.** Containing 30,000 subjects. Cloth, 2s. 6d. ; half-roxburgh, 4s.
**David Balfour, The Adventures of.** By R. L. STEVENSON. Illustrated. Two Vols. 6s. each.
    Part 1.—**Kidnapped.**     Part 2.—**Catriona.**
**Defoe, Daniel, The Life of.** By THOMAS WRIGHT. Illustrated, 21s.
**Diet and Cookery for Common Ailments.** By a Fellow of the Royal College of Physicians, and PHYLLIS BROWNE. 5s.
**Dog, Illustrated Book of the.** By VERO SHAW, B.A. With 28 Coloured Plates. Cloth bevelled, 35s. ; half-morocco, 45s.
**Domestic Dictionary, The.** Illustrated. Cloth, 7s. 6d.
**Doré Bible, The.** With 200 Full-page Illustrations by DORÉ. 15s.
**Doré Don Quixote, The.** With about 400 Illustrations by GUSTAVE DORÉ. *Cheap Edition.* Bevelled boards, gilt edges, 10s. 6d.
**Doré Gallery, The.** With 250 Illustrations by DORÉ. 4to, 42s.
**Doré's Dante's Inferno.** Illustrated by GUSTAVE DORÉ. With Preface by A. J. BUTLER. Cloth gilt or buckram, 7s. 6d.
**Doré's Dante's Purgatory and Paradise.** Illustrated by GUSTAVE DORÉ. *Cheap Edition,* 7s. 6d.
**Doré's Milton's Paradise Lost.** Illustrated by DORÉ. 4to, 21s. *Popular Edition.* Cloth gilt or buckram gilt, 7s. 6d.
**Dorset, Old.** Chapters in the History of the County. By H. J. MOULE, M.A. 10s. 6d.
**Dressmaking, Modern, The Elements of.** By J. E. DAVIS. Illd. 2s.
**Earth, Our, and its Story.** By Dr. ROBERT BROWN, F.L.S. With Coloured Plates and numerous Wood Engravings. Three Vols. 9s. each.
**Edinburgh, Old and New.** With 600 Illustrations. Three Vols. 9s. each.
**Egypt: Descriptive, Historical, and Picturesque.** By Prof. G. EBERS. With 800 Original Engravings. *Popular Edition.* In Two Vols. 42s.
**Electric Current, The.** How Produced and How Used. By R. MULLINEUX WALMSLEY, D.Sc., etc. Illustrated. 10s. 6d.
**Electricity in the Service of Man.** Illustrated. *New and Revised Edition.* 10s. 6d.
**Electricity, Practical.** By Prof. W. E. AYRTON. 7s. 6d.
**Encyclopædic Dictionary, The.** In Fourteen Divisional Vols., 10s. 6d. each ; or Seven Vols., half-morocco, 21s. each ; half-russia, 25s.
**England, Cassell's Illustrated History of.** With upwards of 2,000 Illustrations. *Revised Edition.* Complete in Eight Vols., 9s. each; cloth gilt, and embossed gilt top and headbanded, £4 net the set.

*Selections from Cassell & Company's Publications.*

---

**English Dictionary, Cassell's.** Giving definitions of more than 100,000 Words and Phrases. *Superior Edition*, 5s. *Cheap Edition*, 3s. 6d.
**English Literature, Library of.** By Prof. HENRY MORLEY. Complete in Five Vols., 7s. 6d. each.
**English Literature, The Dictionary of.** By W. DAVENPORT ADAMS. *Cheap Edition*, 7s. 6d.
**English Literature, Morley's First Sketch of.** *Revised Edition*, 7s. 6d.
**English Literature, The Story of.** By ANNA BUCKLAND. 3s. 6d.
**English Writers.** By Prof. HENRY MORLEY. Vols. I. to XI. 5s. each.
**Etiquette of Good Society.** *New Edition*. Edited and Revised by LADY COLIN CAMPBELL. 1s.; cloth, 1s. 6d.
**Fairway Island.** By HORACE HUTCHINSON. *Cheap Edition*. 3s. 6d.
**Fairy Tales Far and Near.** Re-told by Q. Illustrated. 3s. 6d.
**Fiction, Cassell's Popular Library of.** 3s. 6d. each.

| | |
|---|---|
| THE AVENGER OF BLOOD. By J. MACLAREN COBBAN. | THE SNARE OF THE FOWLER. By Mrs. ALEXANDER. |
| A MODERN DICK WHITTINGTON. By JAMES PAYN. | "LA BELLA" AND OTHERS. By EGERTON CASTLE. |
| THE MAN IN BLACK. By STANLEY WEYMAN. | LEONA. By Mrs. MOLESWORTH. |
| A BLOT OF INK. Translated by Q. and PAUL M. FRANCKE. | FOURTEEN TO ONE, ETC. By ELIZABETH STUART PHELPS. |
| THE MEDICINE LADY. By L. T. MEADE. | FATHER STAFFORD. By ANTHONY HOPE. |
| OUT OF THE JAWS OF DEATH. By FRANK BARRETT. | DR. DUMÁNY'S WIFE. By MAURUS JÓKAI. |
| | THE DOINGS OF RAFFLES HAW. By CONAN DOYLE. |

**Field Naturalist's Handbook, The.** By the Revs. J. G. WOOD and THEODORE WOOD. *Cheap Edition*. 2s. 6d.
**Figuier's Popular Scientific Works.** With Several Hundred Illustrations in each. Newly Revised and Corrected. 3s. 6d. each.
THE HUMAN RACE. MAMMALIA. OCEAN WORLD.
THE INSECT WORLD. REPTILES AND BIRDS.
WORLD BEFORE THE DELUGE. THE VEGETABLE WORLD.
**Flora's Feast.** A Masque of Flowers. Penned and Pictured by WALTER CRANE. With 40 Pages in Colours. 5s.
**Football, The Rugby Union Game.** Edited by REV. F. MARSHALL. Illustrated. *New and Enlarged Edition*. 7s. 6d.
**For Glory and Renown.** By D. H. PARRY. Illustrated. 5s.
**France, From the Memoirs of a Minister of.** By STANLEY WEYMAN. 6s.
**Franco-German War, Cassell's History of the.** Complete in Two Vols. Containing about 500 Illustrations. 9s. each.
**Free Lance in a Far Land, A.** By HERBERT COMPTON. 6s.
**Garden Flowers, Familiar.** By SHIRLEY HIBBERD. With Coloured Plates by F. E. HULME, F.L.S. Complete in Five Series. 12s. 6d. each.
**Gardening, Cassell's Popular.** Illustrated. Four Vols. 5s. each.
**Gazetteer of Great Britain and Ireland, Cassell's.** Illustrated. Vols. I. and II. 7s. 6d. each.
**Gladstone, William Ewart, The People's Life of.** Illustrated. 1s.
**Gleanings from Popular Authors.** Two Vols. With Original Illustrations. 4to, 9s. each. Two Vols. in One, 15s.
**Gulliver's Travels.** With 88 Engravings by MORTEN. *Cheap Edition*. Cloth, 3s. 6d.; cloth gilt, 5s.
**Gun and its Development, The.** By W. W. GREENER. With 500 Illustrations. 10s. 6d.
**Heavens, The Story of the.** By Sir ROBERT STAWELL BALL, LL.D., F.R.S., F.R.A.S. With Coloured Plates. *Popular Edition*. 12s. 6d.
**Heroes of Britain in Peace and War.** With 300 Original Illustrations. Two Vols., 3s. 6d. each; or One Vol., 7s. 6d.
**Highway of Sorrow, The.** By HESBA STRETTON and \*\*\*\*\*\*\*\* 6s.

*Selections from Cassell & Company's Publications.*

---

**Hispaniola Plate (1683-1893).** By JOHN BLOUNDELLE-BURTON. 6s.
**Historic Houses of the United Kingdom.** Profusely Illustrated. 10s. 6d.
**History, A Foot-note to.** Eight Years of Trouble in Samoa. By ROBERT LOUIS STEVENSON. 6s.
**Home Life of the Ancient Greeks, The.** Translated by ALICE ZIMMERN. Illustrated. *Cheap Edition.* 5s.
**Horse, The Book of the.** By SAMUEL SIDNEY. With 17 Full-page Collotype Plates of Celebrated Horses of the Day, and numerous other Illustrations. Cloth, 15s.
**Horses and Dogs.** By O. EERELMAN. With Descriptive Text. Translated from the Dutch by CLARA BELL. With Photogravure Frontispiece, 12 Exquisite Collotypes, and several full page and other engravings in the text. 25s. net.
**Houghton, Lord:** The Life, Letters, and Friendships of Richard Monckton Milnes, First Lord Houghton. By Sir WEMYSS REID. In Two Vols., with Two Portraits. 32s.
**Household, Cassell's Book of the.** Complete in Four Vols. 5s. each. Four Vols. in Two, half-morocco, 25s.
**Hygiene and Public Health.** By B. ARTHUR WHITELEGGE, M.D. 7s. 6d.
**Impregnable City, The.** By MAX PEMBERTON. 6s.
**India, Cassell's History of.** By JAMES GRANT. With about 400 Illustrations. Two Vols., 9s. each. One Vol., 15s.
**Iron Pirate, The.** By MAX PEMBERTON. Illustrated. 5s.
**Island Nights' Entertainments.** By R. L. STEVENSON. Illustrated. 6s.
**Kennel Guide, The Practical.** By Dr. GORDON STABLES. 1s.
**Khiva, A Ride to.** By Col. FRED BURNABY. *New Edition.* With Portrait and Seven Illustrations. 3s. 6d.
**King George, In the Days of.** By COL. PERCY GROVES. Illd. 1s. 6d.
**King's Hussar, A.** Edited by HERBERT COMPTON. 6s.
**Ladies' Physician, The.** By a London Physician. *Cheap Edition, Revised and Enlarged.* 3s. 6d.
**Lady Biddy Fane, The Admirable.** By FRANK BARRETT. *New Edition.* With 12 Full-page Illustrations. 6s.
**Lady's Dressing-room, The.** Translated from the French of BARONESS STAFFE by LADY COLIN CAMPBELL. 3s. 6d.
**Letters, the Highway of, and its Echoes of Famous Footsteps.** By THOMAS ARCHER. Illustrated. 10s. 6d.
**Letts's Diaries and other Time-saving Publications** published exclusively by CASSELL & COMPANY. (*A list free on application.*)
**'Lisbeth.** A Novel. By LESLIE KEITH. 6s.
**List, ye Landsmen!** By W. CLARK RUSSELL. 6s.
**Little Minister, The.** By J. M. BARRIE. *Illustrated Edition.* 6s.
**Little Squire, The.** By Mrs. HENRY DE LA PASTURE. 3s. 6d.
**Llollandllaff Legends, The.** By LOUIS LLOLLANDLLAFF. 1s.; cloth, 2s.
**Lobengula, Three Years With, and Experiences in South Africa.** By J. COOPER-CHADWICK. *Cheap Edition,* 2s. 6d.
**Locomotive Engine, The Biography of a.** By HENRY FRITH. 3s. 6d.
**Loftus, Lord Augustus, The Diplomatic Reminiscences of.** First and Second Series. Two Vols., each with Portrait, 32s. each Series.
**London, Greater.** By EDWARD WALFORD. Two Vols. With about 400 Illustrations. 9s. each.
**London, Old and New.** Six Vols., each containing about 200 Illustrations and Maps. Cloth, 9s. each.
**Lost on Du Corrig; or, 'Twixt Earth and Ocean.** By STANDISH O'GRADY. With 8 Full-page Illustrations. 5s.
**Medicine, Manuals for Students of.** (*A List forwarded post free.*)
**Modern Europe, A History of.** By C. A. FYFFE, M.A. *Cheap Edition in One Volume,* 10s. 6d. Library Edition. Illustrated. 3 Vols., 7s. 6d. each.
**Mount Desolation.** An Australian Romance. By W. CARLTON DAWE. *Cheap Edition.* 3s. 6d.

*Selections from Cassell & Company's Publications.*

**Music, Illustrated History of.** By EMIL NAUMANN. Edited by the Rev. Sir F. A. GORE OUSELEY, Bart. Illustrated. Two Vols. 31s. 6d.
**National Library, Cassell's.** In 214 Volumes. Paper covers, 3d.; cloth, 6d. (*A Complete List of the Volumes post free on application.*)
**Natural History, Cassell's Concise.** By E. PERCEVAL WRIGHT, M.A., M.D., F.L.S. With several Hundred Illustrations. 7s. 6d.
**Natural History, Cassell's New.** Edited by Prof. P. MARTIN DUNCAN, M.B., F.R.S., F.G.S. Complete in Six Vols. With about 2,000 Illustrations. Cloth, 9s. each.
**Nature's Wonder Workers.** By KATE R. LOVELL. Illustrated. 3s. 6d.
**New England Boyhood, A.** By EDWARD E. HALE. 3s. 6d.
**New Zealand, Picturesque.** With Preface by Sir W. B. PERCEVAL, K.C.M.G. Illustrated. 6s.
**Nursing for the Home and for the Hospital, A Handbook of.** By CATHERINE J. WOOD. *Cheap Edition.* 1s. 6d.; cloth, 2s.
**Nursing of Sick Children, A Handbook for the.** By CATHERINE J. WOOD. 2s. 6d.
**Ohio, The New.** A Story of East and West. By EDWARD E. HALE. 6s.
**Oil Painting, A Manual of.** By the Hon. JOHN COLLIER. 2s. 6d.
**Old Maids and Young.** By E. D'ESTERRE KEELING. 6s.
**Old Boy's Yarns, An.** By HAROLD AVERY. With 8 Plates. 3s. 6d.
**Our Own Country.** Six Vols. With 1,200 Illustrations. 7s. 6d. each.
**Painting, The English School of.** *Cheap Edition.* 3s. 6d.
**Painting, Practical Guides to.** With Coloured Plates:—

| | |
|---|---|
| MARINE PAINTING. 5s. | WATER-COLOUR PAINTING. 5s. |
| ANIMAL PAINTING. 5s. | NEUTRAL TINT. 5s. |
| CHINA PAINTING. 5s. | SEPIA, in Two Vols., 3s. each; or |
| FIGURE PAINTING. 7s. 6d. | in One Vol., 5s. |
| ELEMENTARY FLOWER PAINTING. 3s. | FLOWERS, AND HOW TO PAINT THEM. 5s. |

**Paris, Old and New.** A Narrative of its History, Its People, and i's Places. By H. SUTHERLAND EDWARDS. Profusely Illustrated. Complete in Two Vols., 9s. each; or gilt edges, 10s. 6d. each.
**Peoples of the World, The.** In Six Vols. By Dr. ROBERT BROWN. Illustrated. 7s. 6d. each.
**Photography for Amateurs.** By T. C. HEPWORTH. *Enlarged and Revised Edition.* Illustrated. 1s.; or cloth, 1s. 6d.
**Phrase and Fable, Dr. Brewer's Dictionary of.** Giving the Derivation, Source, or Origin of Common Phrases, Allusions, and Words that have a Tale to Tell. *Entirely New and Greatly Enlarged Edition.* 10s. 6d.
**Picturesque America.** Complete in Four Vols., with 48 Exquisite Steel Plates and about 800 Original Wood Engravings. £2 2s. each. *Popular Edition*, Vols. I. & II., 18s. each. [the Set.
**Picturesque Canada.** With 600 Original Illustrations. Two Vols. £6 6s.
**Picturesque Europe.** Complete in Five Vols. Each containing 13 Exquisite Steel Plates, from Original Drawings, and nearly 200 Original Illustrations. Cloth, £21; half-morocco, £31 10s.; morocco gilt, £52 10s. POPULAR EDITION. In Five Vols., 18s. each.
**Picturesque Mediterranean, The.** With Magnificent Original Illustrations by the leading Artists of the Day. Complete in Two Vo's. £2 2s. each.
**Pigeon Keeper, The Practical.** By LEWIS WRIGHT. Illustrated. 3s. 6d.
**Pigeons, Fulton's Book of.** Edited by LEWIS WRIGHT. Revised, Enlarged and supplemented by the Rev. W. F. LUMLEY. With 50 Full page Illustrations. *Popular Edition.* In One Vol., 10s. 6d.
- **Planet, The Story of Our.** By T. G. BONNEY, D.Sc., LL.D., F.R.S., F.S.A., F.G.S. With Coloured Plates and Maps and about 100 Illustrations. 31s. 6d.
**Pocket Library, Cassell's.** Cloth, 1s. 4d. each.
  A King's Diary. By PERCY WHITE.
  A While B by. By JAMES WELSH.
  The Little Huguenot. By MAX PEMBERTON
  A Whirl Asunder. By GERTRUDE ATHERTO

*Selections from Cassell & Company's Publications.*

**Poems, Aubrey de Vere's.** A Selection. Edited by J. DENNIS. 3s. 6d.
**Poets, Cassell's Miniature Library of the.** Price 1s. each Vol.
**Pomona's Travels.** By FRANK R. STOCKTON. Illustrated. 5s.
**Portrait Gallery, The Cabinet.** Complete in Five Series, each containing 36 Cabinet Photographs of Eminent Men and Women. 15s. each.
**Portrait Gallery, Cassell's Universal.** Containing 240 Portraits of Celebrated Men and Women of the Day. With brief Memoirs and *facsimile* Autographs. Cloth, 6s.
**Poultry Keeper, The Practical.** By L. WRIGHT. Illustrated. 3s. 6d.
**Poultry, The Book of.** By LEWIS WRIGHT. *Popular Edition.* 10s. 6d.
**Poultry, The Illustrated Book of.** By LEWIS WRIGHT. With Fifty Coloured Plates. *New and Revised Edition.* Cloth, gilt edges (*Price on application*). Half-morocco, £2 2s.
**Prison Princess, A.** By Major ARTHUR GRIFFITHS. 6s.
**"Punch," The History of.** By M. H. SPIELMANN. With upwards of 160 Illustrations, Portraits, and Facsimiles. Cloth, 16s.; *Large Paper Edition*, £2 2s. net.
**Q's Works, Uniform Edition of.** 5s. each.

Dead Man's Rock. | The Astonishing History of Troy Town.
The Splendid Spur. | "I Saw Three Ships," and other Winter's Tales.
The Blue Pavilions. | Noughts and Crosses.
The Delectable Duchy.

**Queen Summer;** or, The Tourney of the Lily and the Rose. With Forty Pages of Designs in Colours by WALTER CRANE. 6s.
**Queen, The People's Life of their.** By Rev. E. J. HARDY, M.A. 1s.
**Queen Victoria, The Life and Times of.** By ROBERT WILSON. Complete in Two Vols. With numerous Illustrations. 9s. each.
**Queen's Scarlet, The.** By G. MANVILLE FENN. Illustrated. 5s.
**Rabbit-Keeper, The Practical.** By CUNICULUS. Illustrated. 3s. 6d.
**Railways, Our.** Their Origin, Development, Incident, and Romance. By JOHN PENDLETON. Illustrated. 2 Vols., 24s.
**Railway Guides, Official Illustrated.** With Illustrations, Maps, &c. Price 1s. each; or in cloth, 2s. each.

LONDON AND NORTH WESTERN RAILWAY. | GREAT EASTERN RAILWAY.
GREAT WESTERN RAILWAY. | LONDON AND SOUTH WESTERN RAILWAY.
MIDLAND RAILWAY. | LONDON, BRIGHTON AND SOUTH COAST RAILWAY.
GREAT NORTHERN RAILWAY. | SOUTH-EASTERN RAILWAY.

**Railway Guides, Official Illustrated.** Abridged and Popular Editions. Paper covers, 3d. each.

GREAT EASTERN RAILWAY. | LONDON AND SOUTH WESTERN RAILWAY.
LONDON AND NORTH WESTERN RAILWAY. |

**Railway Library, Cassell's.** Crown 8vo, boards, 2s. each. (*A List of the Vols. post free on application.*)
**Red Terror, The.** A Story of the Paris Commune. By EDWARD KING. Illustrated. 3s. 6d.
**Rivers of Great Britain:** Descriptive, Historical, Pictorial.
THE ROYAL RIVER: The Thames, from Source to Sea. 16s.
RIVERS OF THE EAST COAST. *Popular Edition*, 16s.
**Robinson Crusoe, Cassell's New Fine-Art Edition of.** 7s. 6d.
**Romance, The World of.** Illustrated. Cloth, 9s.
**Royal Academy Pictures, 1895.** With upwards of 200 magnificent reproductions of Pictures in the Royal Academy of 1895. 7s. 6d.
**Russo-Turkish War, Cassell's History of.** With about 500 Illustrations. Two Vols. 9s. each.
**Sala, George Augustus, The Life and Adventures of.** By Himself. In Two Vols., demy 8vo, cloth, 32s.
**Saturday Journal, Cassell's.** Yearly Volume, cloth, 7s. 6d.

*Selections from Cassell & Company's Publications.*

**Science Series, The Century.** Consisting of Biographies of Eminent Scientific Men of the present Century. Edited by Sir HENRY ROSCOE, D.C.L., F.R.S. Crown 8vo, 3s. 6d. each.
**John Dalton and the Rise of Modern Chemistry.** By Sir HENRY E. ROSCOE, F.R.S.
**Major Rennell, F.R.S., and the Rise of English Geography.** By CLEMENTS R. MARKHAM, C.B., F.R.S., President of the Royal Geographical Society.
**Justus Von Liebig: His Life and Work.** By W. A. SHENSTONE, F.I.C.
**The Herschels and Modern Astronomy.** By MISS AGNES M. CLERKE.
**Charles Lyell: His Life and Work.** By Professor T. G. BONNEY, F.R.S.
**Science for All.** Edited by Dr. ROBERT BROWN. Five Vols. 9s. each.
**Scotland, Picturesque and Traditional.** A Pilgrimage with Staff and Knapsack. By G. E. EYRE-TODD. 6s.
**Sea, The Story of the.** An Entirely New and Original Work. Edited by Q. Illustrated. Vol. I. 9s.
**Sea Wolves, The.** By MAX PEMBERTON. Illustrated. 6s.
**Shadow of a Song, The.** A Novel. By CECIL HARLEY. 5s.
**Shaftesbury, The Seventh Earl of, K.G.,** The Life and Work of. By EDWIN HODDER. *Cheap Edition.* 3s. 6d.
**Shakespeare, The Plays of.** Edited by Professor HENRY MORLEY. Complete in Thirteen Vols., cloth, 21s.; half-morocco, cloth sides, 42s.
**Shakespeare, Cassell's Quarto Edition.** Containing about 600 Illustrations by H. C. SELOUS. Complete in Three Vols., cloth gilt, £3 3s.
**Shakespeare, The England of.** *New Edition.* By E. GOADBY. With Full-page Illustrations. 2s. 6d.
**Shakspere's Works.** *Édition de Luxe.*
"**King Henry VIII.**" Illustrated by SIR JAMES LINTON, P.R.I. (*Price on application.*)
"**Othello.**" Illustrated by FRANK DICKSEE, R.A. £3 10s.
"**King Henry IV.**" Illustrated by EDUARD GRÜTZNER. £3 10s.
"**As You Like It.**" Illustrated by EMILE BAYARD. £3 10s.
**Shakspere, The Leopold.** With 400 Illustrations. *Cheap Edition.* 3s. 6d. Cloth gilt, gilt edges, 5s.; Roxburgh, 7s. 6d.
**Shakspere, The Royal.** With Steel Plates and Wood Engravings. Three Vols. 15s. each.
**Sketches, The Art of Making and Using.** From the French of G. FRAIPONT. By CLARA BELL. With 50 Illustrations. 2s. 6d.
**Smuggling Days and Smuggling Ways.** By Commander the Hon. HENRY N. SHORE, R.N. With numerous Illustrations. 7s. 6d.
**Social England.** A Record of the Progress of the People. By various writers. Edited by H. D. TRAILL, D.C.L. Vols. I., II., & III., 15s. each. Vol. IV., 17s.
**Social Welfare, Subjects of.** By Rt. Hon. LORD PLAYFAIR, K.C.B. 7s. 6d.
**Sports and Pastimes, Cassell's Complete Book of.** *Cheap Edition.* With more than 900 Illustrations. Medium 8vo, 992 pages, cloth, 3s. 6d.
**Squire, The.** By Mrs. PARR. *Popular Edition.* 6s.
**Standishs of High Acre, The.** A Novel. By GILBERT SHELDON. Two Vols. 21s.
**Star-Land.** By Sir R. S. BALL, LL.D., &c. Illustrated. 6s.
**Statesmen, Past and Future.** 6s.
**Story of Francis Cludde, The.** By STANLEY J. WEYMAN. 6s.
**Story Poems.** For Young and Old. Edited by E. DAVENPORT. 3s. 6d.
**Sun, The.** By Sir ROBERT STAWELL BALL, LL.D., F.R.S., F.R.A.S. With Eight Coloured Plates and other Illustrations. 21s.
**Sunshine Series, Cassell's.** 1s. each.
(*A List of the Volumes post free on application.*)
**Thackeray in America, With.** By EYRE CROWE, A.R.A. Ill. 10s. 6d.
**The "Treasure Island" Series.** *Illustrated Edition.* 3s. 6d. each.

| | |
|---|---|
| **Treasure Island.** By ROBERT LOUIS STEVENSON. | **The Black Arrow.** By ROBERT LOUIS STEVENSON. |
| **The Master of Ballantrae.** By ROBERT LOUIS STEVENSON. | **King Solomon's Mines.** By H. RIDER HAGGARD. |

*Selections from Cassell & Company's Publications.*

**Things I have Seen and People I have Known.** By G. A. SALA. With Portrait and Autograph. 2 Vols. 21s.
**Tidal Thames, The.** By GRANT ALLEN. With India Proof Impressions of Twenty magnificent Full-page Photogravure Plates, and with many other Illustrations in the Text after Original Drawings by W. L. WYLLIE, A.R.A. Half morocco. £5 15s. 6d.
**Tiny Luttrell.** By E. W. HORNUNG. *Popular Edition.* 6s.
**To Punish the Czar: a Story of the Crimea.** By HORACE HUTCHINSON. Illustrated. 3s. 6d.
**Treatment, The Year-Book of, for 1896.** A Critical Review for Practitioners of Medicine and Surgery. *Twelfth Year of Issue.* 7s. 6d.
**Trees, Familiar.** By G. S. BOULGER, F.L.S. Two Series. With 40 full-page Coloured Plates by W. H. J. BOOT. 12s. 6d. each.
**Tuxter's Little Maid.** By G. B. BURGIN. 6s.
**"Unicode": the Universal Telegraphic Phrase Book.** *Desk or Pocket Edition.* 2s. 6d.
**United States, Cassell's History of the.** By EDMUND OLLIER. With 600 Illustrations. Three Vols. 9s. each.
**Universal History, Cassell's Illustrated.** Four Vols. 9s. each.
**Vision of Saints, A.** By Sir LEWIS MORRIS. With 20 Full-page Illustrations. Crown 4to, cloth, 10s. 6d. *Non-illustrated Edition*, 6s.
**Wandering Heath.** Short Stories. By Q. 6s.
**War and Peace, Memories and Studies of.** By ARCHIBALD FORBES. 16s.
**Westminster Abbey, Annals of.** By E. T. BRADLEY (Mrs. A. MURRAY SMITH). Illustrated. With a Preface by Dean BRADLEY. 63s.
**White Shield, The.** By BERTRAM MITFORD. 6s.
**Wild Birds, Familiar.** By W. SWAYSLAND. Four Series. With 40 Coloured Plates in each. (Sold in sets only; price on application.)
**Wild Flowers, Familiar.** By F. E. HULME, F.L.S., F.S.A. Five Series. With 40 Coloured Plates in each. (In sets only; price on application.)
**Wild Flowers Collecting Book.** In Six Parts, 4d. each.
**Wild Flowers Drawing and Painting Book.** In Six Parts, 4d. each.
**Windsor Castle, The Governor's Guide to.** By the Most Noble the MARQUIS OF LORNE, K.T. Profusely Illustrated. Limp Cloth, 1s. Cloth boards, gilt edges, 2s.
**Wit and Humour, Cassell's New World of.** With New Pictures and New Text. 6s.
**With Claymore and Bayonet.** By Col. PERCY GROVES. Illd. 5s.
**Wood, Rev. J. G., Life of the.** By the Rev. THEODORE WOOD. Extra crown 8vo, cloth. *Cheap Edition.* 3s. 6d.
**Work.** The Illustrated Weekly Journal for Mechanics. Vol. IX., 4s.
**"Work" Handbooks.** Practical Manuals prepared *under the direction of* PAUL N. HASLUCK, Editor of *Work.* Illustrated. 1s. each.
**World Beneath the Waters, A.** By Rev. GERARD BANCKS. 3s. 6d.
**World of Wonders.** Two Vols. With 400 Illustrations. 7s. 6d. each.
**Wrecker, The.** By R. L. STEVENSON and L. OSBOURNE. Illustrated. 6s.
**Yule Tide.** Cassell's Christmas Annual. 1s.

*ILLUSTRATED MAGAZINES.*

*The Quiver.* Monthly, 6d.
*Cassell's Family Magazine.* Monthly, 6d.
*"Little Folks" Magazine.* Monthly, 6d.
*The Magazine of Art.* Monthly, 1s. 4d.
*"Chums."* Illustrated Paper for Boys. Weekly, 1d.; Monthly, 6d.
*Cassell's Saturday Journal.* Weekly, 1d.; Monthly, 6d.
*Work.* Weekly, 1d.; Monthly, 6d.
*Cottage Gardening.* Weekly, ½d.; Monthly, 3d.

CASSELL & COMPANY, LIMITED, *Ludgate Hill, London.*

*Selections from Cassell & Company's Publications.*

## Bibles and Religious Works.

**Bible Biographies.** Illustrated. 2s. 6d. each.
The Story of Moses and Joshua. By the Rev. J. TELFORD.
The Story of the Judges. By the Rev. J. WYCLIFFE GEDGE.
The Story of Samuel and Saul. By the Rev. D. C. TOVEY.
The Story of David. By the Rev. J. WILD.
The Story of Joseph. Its Lessons for To-Day. By the Rev. GEORGE BAINTON
The Story of Jesus. In Verse. By J. R. MACDUFF, D.D.

**Bible, Cassell's Illustrated Family.** With 900 Illustrations. Leather, gilt edges, £2 10s.
**Bible Educator, The.** Edited by the Very Rev. Dean PLUMPTRE, D.D. With Illustrations, Maps, &c. Four Vols., cloth, 6s. each.
**Bible Manual, Cassell's Illustrated.** By the Rev. ROBERT HUNTER, LL.D. *Illustrated.* 7s. 6d.
**Bible Student in the British Museum, The.** By the Rev. J. G. KITCHIN, M.A. *New and Revised Edition.* 1s. 4d.
**Biblewomen and Nurses.** Yearly Volume. Illustrated. 3s.
**Bunyan, Cassell's Illustrated.** With 200 Original Illustrations. *Cheap Edition.* 7s. 6d.
**Bunyan's Pilgrim's Progress.** Illustrated throughout. Cloth, 3s. 6d.; cloth gilt, gilt edges, 5s.
**Child's Bible, The.** With 200 Illustrations. 150th *Thousand.* 7s. 6d.
**Child's Life of Christ, The.** With 200 Illustrations. 7s. 6d.
**"Come, ye Children."** Illustrated. By Rev. BENJAMIN WAUGH. 3s. 6d.
**Conquests of the Cross.** Illustrated. In 3 Vols. 9s. each.
**Doré Bible.** With 238 Illustrations by GUSTAVE DORÉ. Small folio, best morocco, gilt edges, £15. *Popular Edition.* With 200 Illustrations. 15s.
**Early Days of Christianity, The.** By the Very Rev. Dean FARRAR, D.D., F.R.S. LIBRARY EDITION. Two Vols., 24s.; morocco, £2 2s. POPULAR EDITION. Complete in One Volume, cloth, 6s.; cloth, gilt edges, 7s. 6d.; Persian morocco, 10s. 6d.; tree-calf, 15s.
**Family Prayer-Book, The.** Edited by Rev. Canon GARBETT, M.A., and Rev. S. MARTIN. With Full page Illustrations. *New Edition.* Cloth, 7s. 6d.
**Gleanings after Harvest.** Studies and Sketches by the Rev. JOHN R. VERNON, M.A. Illustrated. 6s.
**"Graven in the Rock."** By the Rev. Dr. SAMUEL KINNS, F.R.A.S., Author of "Moses and Geology." Illustrated. 12s. 6d.
**"Heart Chords."** A Series of Works by Eminent Divines. Bound in cloth, red edges, One Shilling each.

| | |
|---|---|
| MY BIBLE. By the Right Rev. W. BOYD CARPENTER, Bishop of Ripon. | MY GROWTH IN DIVINE LIFE. By the Rev. Preb. REYNOLDS, M.A. |
| MY FATHER. By the Right Rev. ASHTON OXENDEN, late Bishop of Montreal. | MY SOUL. By the Rev. P. B. POWER, M.A. |
| MY WORK FOR GOD. By the Right Rev. Bishop COTTERILL. | MY HEREAFTER. By the Very Rev. Dean BICKERSTETH. |
| MY OBJECT IN LIFE. By the Very Rev. Dean FARRAR, D.D. | MY WALK WITH GOD. By the Very Rev. Dean MONTGOMERY. |
| MY ASPIRATIONS. By the Rev. G. MATHESON, D.D. | MY AIDS TO THE DIVINE LIFE. By the Very Rev. Dean BOYLE. |
| MY EMOTIONAL LIFE. By the Rev. Preb. CHADWICK, D.D. | MY SOURCES OF STRENGTH. By the Rev. E. E. JENKINS, M.A., Secretary of Wesleyan Missionary Society. |
| MY BODY. By the Rev. Prof. W. G. BLAIKIE, D.D. | |

**Helps to Belief.** A Series of Helpful Manuals on the Religious Difficulties of the Day. Edited by the Rev. TEIGNMOUTH SHORE, M.A., Canon of Worcester. Cloth, 1s. each.

| | |
|---|---|
| CREATION. By Harvey Goodwin, D.D., late Bishop of Carlisle. | PRAYER. By the Rev. Canon Shore, M.A. |
| THE DIVINITY OF OUR LORD. By the Lord Bishop of Derry. | THE ATONEMENT. By William Connor Magee, D.D., Late Archbishop of York. |
| MIRACLES. By the Rev. Brownlow Maitland, M.A. | |

5 B. 8.95

*Selections from Cassell & Company's Publications.*

**Holy Land and the Bible, The.** By the Rev. C. GEIKIE, D.D., LL.D. (Edin.). Two Vols., 24s. *Illustrated Edition*, One Vol., 21s.
**Life of Christ, The.** By the Very Rev. Dean FARRAR, D.D., F.R.S. LIBRARY EDITION. Two Vols. Cloth, 24s.; morocco, 42s. CHEAP ILLUSTRATED EDITION. Cloth, 7s. 6d.; cloth, full gilt, gilt edges, 10s. 6d. POPULAR EDITION (*Revised and Enlarged*), 8vo, cloth, gilt edges, 7s. 6d.; Persian morocco, gilt edges, 10s. 6d.; tree-calf, 15s.
**Moses and Geology**; or, The Harmony of the Bible with Science. By the Rev. SAMUEL KINNS, Ph.D., F.R.A.S. Illustrated. *New Edition*. 10s. 6d.
**My Last Will and Testament.** By HYACINTHE LOYSON (Père Hyacinthe). Translated by FABIAN WARE. 1s.; cloth, 1s. 6d.
**New Light on the Bible and the Holy Land.** By B. T. A. EVETTS, M.A. Illustrated. 21s.
**New Testament Commentary for English Readers, The.** Edited by Bishop ELLICOTT. In Three Volumes. 21s. each. Vol. I.—The Four Gospels. Vol. II.—The Acts, Romans, Corinthians, Galatians. Vol. III.—The remaining Books of the New Testament.
**New Testament Commentary.** Edited by Bishop ELLICOTT. Handy Volume Edition. St. Matthew, 3s. 6d. St. Mark, 3s. St. Luke, 3s. 6d. St. John, 3s. 6d. The Acts of the Apostles, 3s. 6d. Romans, 2s. 6d. Corinthians I. and II., 3s. Galatians, Ephesians, and Philippians, 3s. Colossians, Thessalonians, and Timothy, 3s. Titus, Philemon, Hebrews, and James, 3s. Peter, Jude, and John, 3s. The Revelation, 3s. An Introduction to the New Testament, 3s. 6d.
**Old Testament Commentary for English Readers, The.** Edited by Bishop ELLICOTT. Complete in Five Vols. 21s. each. Vol. I.—Genesis to Numbers. Vol. II.—Deuteronomy to Samuel II. Vol. III.—Kings I. to Esther. Vol. IV.—Job to Isaiah. Vol. V.—Jeremiah to Malachi.
**Old Testament Commentary.** Edited by Bishop ELLICOTT. Handy Volume Edition. Genesis, 3s. 6d. Exodus, 3s. Leviticus, 3s. Numbers, 2s. 6d. Deuteronomy, 2s. 6d.
**Plain Introductions to the Books of the Old Testament.** Edited by Bishop ELLICOTT. 3s. 6d.
**Plain Introductions to the Books of the New Testament.** Edited by Bishop ELLICOTT. 3s. 6d.
**Protestantism, The History of.** By the Rev. J. A. WYLIE, LL.D. Containing upwards of 600 Original Illustrations. Three Vols. 9s. each.
**Quiver Yearly Volume, The.** With about 600 Original Illustrations. 7s. 6d.
**Religion, The Dictionary of.** By the Rev. W. BENHAM, B.D. *Cheap Edition*. 10s. 6d.
**St. George for England**; and other Sermons preached to Children. By the Rev. T. TEIGNMOUTH SHORE, M.A., Canon of Worcester. 5s.
**St. Paul, The Life and Work of.** By the Very Rev. Dean FARRAR, D.D., F.R.S. LIBRARY EDITION. Two Vols., cloth, 24s.; calf, 42s. ILLUSTRATED EDITION, complete in One Volume, with about 300 Illustrations, £1 1s.; morocco, £2 2s. POPULAR EDITION. One Volume, 8vo, cloth, 6s.; cloth, gilt edges, 7s. 6d.; Persian morocco, 10s. 6d.; tree-calf, 15s.
**Shall We Know One Another in Heaven?** By the Rt. Rev. J. C. RYLE, D.D., Bishop of Liverpool. *Cheap Edition*. Paper covers, 6d.
**Searchings in the Silence.** By Rev. GEORGE MATHESON, D.D. 3s. 6d.
**"Sunday," Its Origin, History, and Present Obligation.** By the Ven. Archdeacon HESSEY, D.C.L. *Fifth Edition*. 7s. 6d.
**Twilight of Life, The.** Words of Counsel and Comfort for the Aged. By the Rev. JOHN ELLERTON, M.A. 1s. 6d.

*Selections from Cassell & Company's Publications.*

## Educational Works and Students' Manuals.

**Agricultural Text-Books, Cassell's.** (The "Downton" Series.) Edited by JOHN WRIGHTSON, Professor of Agriculture. Fully Illustrated, 2s. 6d. each.—Farm Crops. By Prof. WRIGHTSON.—Soils and Manures. By J. M. H. MUNRO, D.Sc. (London), F.I.C., F.C.S. —Live Stock. By Prof. WRIGHTSON.
**Alphabet, Cassell's Pictorial.** 3s. 6d.
**Arithmetics, Cassell's "Belle Sauvage."** By GEORGE RICKS, B.Sc. Lond. With Test Cards. (*List on application.*)
**Atlas, Cassell's Popular.** Containing 24 Coloured Maps. 2s. 6d.
**Book-Keeping.** By THEODORE JONES. For Schools, 2s.; cloth, 3s. For the Million, 2s.; cloth, 3s. Books for Jones's System, 2s.
**British Empire Map of the World.** By G. R. PARKIN and J. G. BARTHOLOMEW, F.R.G.S. 25s.
**Chemistry, The Public School.** By J. H. ANDERSON, M.A. 2s. 6d.
**Cookery for Schools.** By LIZZIE HERITAGE. 6d.
**Dulce Domum.** Rhymes and Songs for Children. Edited by JOHN FARMER, Editor of "Gaudeamus," &c. Old Notation and Words, 5s. N.B.—The words of the Songs in "Dulce Domum" (with the Airs both in Tonic Sol-fa and Old Notation) can be had in Two Parts, 6d. each.
**Euclid, Cassell's.** Edited by Prof. WALLACE, M.A. 1s.
**Euclid, The First Four Books of.** *New Edition.* In paper, 6d.; cloth, 9d.
**Experimental Geometry.** By PAUL BERT. Illustrated. 1s. 6d.
**French, Cassell's Lessons in.** *New and Revised Edition.* Parts I. and II., each 2s. 6d.; complete, 4s. 6d. Key, 1s. 6d.
**French-English and English-French Dictionary.** *Entirely New and Enlarged Edition.* Cloth, 3s. 6d.; superior binding, 5s.
**French Reader, Cassell's Public School.** By G. S. CONRAD. 2s. 6d.
**Gaudeamus.** Songs for Colleges and Schools. Edited by JOHN FARMER. 5s. Words only, paper covers, 6d.; cloth, 9d.
**German Dictionary, Cassell's New** (German-English, English-German). *Cheap Edition.* Cloth, 3s. 6d. *Superior Binding*, 5s.
**Hand and Eye Training.** By G. RICKS, B.Sc. 2 Vols., with 16 Coloured Plates in each Vol. Cr. 4to, 6s. each. Cards for Class Use, 5 sets, 1s. each.
**Hand and Eye Training.** By GEORGE RICKS, B.Sc., and JOSEPH VAUGHAN. Illustrated. Vol. I. Designing with Coloured Papers. Vol. II. Cardboard Work. 2s. each. Vol. III. Colour Work and Design, 3s.
**Historical Cartoons, Cassell's Coloured.** Size 45 in. × 35 in., 2s. each. Mounted on canvas and varnished, with rollers, 5s. each.
**Italian Lessons, with Exercises, Cassell's.** Cloth, 3s. 6d.
**Latin Dictionary, Cassell's New.** (Latin-English and English-Latin.) Revised by J. R. V. MARCHANT, M.A., and J. F. CHARLES, B.A. Cloth, 3s. 6d. *Superior Binding*, 5s.
**Latin Primer, The First.** By Prof. POSTGATE. 1s.
**Latin Primer, The New.** By Prof. J. P. POSTGATE. 2s. 6d.
**Latin Prose for Lower Forms.** By M. A. BAYFIELD, M.A. 2s. 6d.
**Laws of Every-Day Life.** By H. O. ARNOLD-FORSTER, M.P. 1s. 6d. *Special Edition* on Green Paper for Persons with Weak Eyesight. 2s.
**Lessons in Our Laws; or, Talks at Broadacre Farm.** By H. F. LESTER, B.A. Parts I. and II., 1s. 6d. each.
**Little Folks' History of England.** Illustrated. 1s. 6d.
**Making of the Home, The.** By Mrs. SAMUEL A. BARNETT. 1s. 6d.
**Marlborough Books:**—Arithmetic Examples, 3s. French Exercises, 3s. 6d. French Grammar, 2s. 6d. German Grammar, 3s. 6d.
**Mechanics and Machine Design, Numerical Examples in Practical.** By R. G. BLAINE, M.E. *New Edition, Revised and Enlarged.* With 79 Illustrations. Cloth, 2s. 6d.
**Mechanics for Young Beginners, A First Book of.** By the Rev. J. G. EASTON, M.A. 4s. 6d.

*Selections from Cassell & Company's Publications.*

---

**Natural History Coloured Wall Sheets, Cassell's** New. 17 Subjects. Size 39 by 31 in. Mounted on rollers and varnished. 3s. each.
**Object Lessons from Nature.** By Prof. L. C. MIALL, F.L.S. Fully Illustrated. *New and Enlarged Edition.* Two Vols., 1s. 6d. each.
**Physiology for Schools.** By A. T. SCHOFIELD, M.D., M.R.C.S., &c. Illustrated. Cloth, 1s. 9d.; Three Parts, paper covers, 5d. each; or cloth limp, 6d. each.
**Poetry Readers, Cassell's New.** Illustrated. 12 Books, 1d. each; or complete in one Vol., cloth, 1s. 6d.
**Popular Educator, Cassell's NEW.** With Revised Text, New Maps, New Coloured Plates, New Type, &c. In 8 Vols., 5s. each; or in Four Vols., half-morocco, 50s. the set.
**Readers, Cassell's " Belle Sauvage."** An entirely New Series. Fully Illustrated. Strongly bound in cloth. (*List on application.*)
**Readers, Cassell's " Higher Class."** (*List on application.*)
**Readers, Cassell's Readable.** Illustrated. (*List on application.*)
**Readers for Infant Schools, Coloured.** Three Books. 4d. each.
**Reader, The Citizen.** By H. O. ARNOLD-FORSTER, M.P. Illustrated. 1s. 6d. Also a *Scottish Edition*, cloth, 1s. 6d.
**Reader, The Temperance.** By Rev. J. DENNIS HIRD. 1s. 6d.
**Readers, Geographical, Cassell's New.** With numerous Illustrations. (*List on application.*)
**Readers, The " Modern School" Geographical.** (*List on application.*)
**Readers, The " Modern School."** Illustrated. (*List on application.*)
**Reckoning, Howard's Art of.** By C. FRUSHER HOWARD. Paper covers, 1s.; cloth, 2s. *New Edition*, 5s.
**Round the Empire.** By G. R. PARKIN. Fully Illustrated. 1s. 6d.
**Science Applied to Work.** By J. A. BOWER. 1s.
**Science of Everyday Life** By J. A. BOWER. Illustrated. 1s.
**Shade from Models, Common Objects, and Casts of Ornament, How to.** By W. E. SPARKES. With 25 Plates by the Author. 3s.
**Shakspere's Plays for School Use.** 9 Books. Illustrated. 6d. each.
**Spelling, A Complete Manual of.** By J. D. MORELL, LL.D. 1s.
**Technical Manuals, Cassell's.** Illustrated throughout:—
Handrailing and Staircasing, 3s. 6d.—Bricklayers, Drawing for, 3s.—Building Construction, 2s. — Cabinet-Makers, Drawing for, 3s. — Carpenters and Joiners, Drawing for, 3s. 6d.—Gothic Stonework, 3s.—Linear Drawing and Practical Geometry, 2s.—Linear Drawing and Projection. The Two Vols. in One, 3s. 6d.—Machinists and Engineers, Drawing for, 4s. 6d.—Metal-Plate Workers, Drawing for, 3s.—Model Drawing, 3s.—Orthographical and Isometrical Projection, 2s.—Practical Perspective, 3s.—Stonemasons, Drawing for, 3s.—Applied Mechanics, by Sir R. S. Ball, LL.D., 2s.—Systematic Drawing and Shading, 2s.
**Technical Educator, Cassell's New.** With Coloured Plates and Engravings. Complete in Six Volumes, 5s. each.
**Technology, Manuals of.** Edited by Prof. AYRTON, F.R.S., and RICHARD WORMELL, D.Sc., M.A. Illustrated throughout:—
The Dyeing of Textile Fabrics, by Prof. Hummel, 5s.—Watch and Clock Making, by D. Glasgow, Vice-President of the British Horological Institute, 4s. 6d.—Steel and Iron, by Prof. W. H. Greenwood, F.C.S., M.I.C.E., &c., 5s.—Spinning Woollen and Worsted, by W. S. B. McLaren, M.P., 4s. 6d.—Design in Textile Fabrics, by T. R. Ashenhurst, 4s. 6d.—Practical Mechanics, by Prof. Perry, M.E., 3s. 6d.—Cutting Tools Worked by Hand and Machine, by Prof. Smith, 3s. 6d.
**Things New and Old; or, Stories from English History.** By H. O. ARNOLD-FORSTER, M.P. Fully Illustrated, and strongly bound in Cloth. Standards I. & II., 9d. each; Standard III., 1s.; Standard IV., 1s. 3d.; Standards V., VI., & VII., 1s. 6d. each.
**This World of Ours.** By H. O. ARNOLD-FORSTER, M.P. Illustrated. 3s. 6d.

*Selections from Cassell & Company's Publications.*

---

## Books for Young People.

**"Little Folks" Half-Yearly Volume.** Containing 432 4to pages, with about 200 Illustrations, and Pictures in Colour. Boards, 3s. 6d.; cloth, 5s.

**Bo-Peep.** A Book for the Little Ones. With Original Stories and Verses. Illustrated throughout. Yearly Volume. Boards, 2s. 6d.; cloth, 3s. 6d.

**Beneath the Banner.** Being Narratives of Noble Lives and Brave Deeds. By F. J. CROSS. Illustrated. Limp cloth, 1s. Cloth gilt, 2s.

**Good Morning! Good Night!** By F. J. CROSS. Illustrated. Limp cloth, 1s., or cloth boards, gilt lettered, 2s.

**Five Stars in a Little Pool.** By EDITH CARRINGTON. Illustrated. 6s.

**The Cost of a Mistake.** By SARAH PITT. Illustrated. *New Edition.* 2s 6d.

**Beyond the Blue Mountains.** By L. T. MEADE. 5s.

**The Peep of Day.** *Cassell's Illustrated Edition.* 2s. 6d.

**Maggie Steele's Diary.** By E. A. DILLWYN. 2s. 6d.

**A Book of Merry Tales.** By MAGGIE BROWNE, "SHEILA," ISABEL WILSON, and C. L. MATÉAUX. Illustrated. 3s. 6d.

**A Sunday Story-Book.** By MAGGIE BROWNE, SAM BROWNE, and AUNT ETHEL. Illustrated. 3s. 6d.

**A Bundle of Tales.** By MAGGIE BROWNE (Author of "Wanted—a King," &c.), SAM BROWNE, and AUNT ETHEL. 3s. 6d.

**Pleasant Work for Busy Fingers.** By MAGGIE BROWNE. Illustrated. 5s.

**Born a King.** By FRANCES and MARY ARNOLD-FORSTER. (The Life of Alfonso XIII., the Boy King of Spain.) Illustrated. 1s.

**Cassell's Pictorial Scrap Book.** Six Vols. 3s. 6d. each.

**Schoolroom and Home Theatricals.** By ARTHUR WAUGH. Illustrated. *New Edition.* Paper, 1s. Cloth, 1s. 6d.

**Magic at Home.** By Prof. HOFFMAN. Illustrated. Cloth gilt, 3s. 6d.

**Little Mother Bunch.** By Mrs. MOLESWORTH. Illustrated. *New Edition.* Cloth. 2s. 6d.

**Heroes of Every-day Life.** By LAURA LANE. With about 20 Full-page Illustrations. Cloth. 2s. 6d.

**Bob Lovell's Career.** By EDWARD S. ELLIS. 5s.

**Books for Young People.** *Cheap Edition.* Illustrated. Cloth gilt, 3s. 6d. each.

The Champion of Odin; or, Viking Life in the Days of Old. By J. Fred. Hodgetts.
Bound by a Spell; or, The Hunted Witch of the Forest. By the Hon. Mrs. Greene.
Under Bayard's Banner. By Henry Frith.

**Books for Young People.** Illustrated. 3s. 6d. each.

Told Out of School. By A. J. Daniels.
Red Rose and Tiger Lily. By L. T. Meade.
The Romance of Invention. By James Burnley.
*Bashful Fifteen. By L. T. Meade.
*The White House at Inch Gow. By Mrs. Pitt.
*A Sweet Girl Graduate. By L. T. Meade.
The King's Command: A Story for Girls. By Maggie Symington.
*The Palace Beautiful. By L. T. Meade.
*Polly: A New-Fashioned Girl. By L. T. Meade.
"Follow My Leader." By Talbot Baines Reed.
*A World of Girls: The Story of a School. By L. T. Meade.
Lost among White Africans. By David Ker.
For Fortune and Glory: A Story of the Soudan War. By Lewis Hough.

*Also procurable in superior binding, 5s. each.

*Selections from Cassell & Company's Publications.*

---

**"Peeps Abroad" Library.** *Cheap Editions.* Gilt edges, 2s. 6d. each.

Rambles Round London. By C. L. Matéaux. Illustrated.
Around and About Old England. By C. L. Matéaux. Illustrated.
Paws and Claws. By one of the Authors of "Poems written for a Child." Illustrated.
Decisive Events in History. By Thomas Archer. With Original Illustrations.
The True Robinson Crusoes. Cloth gilt.

Peeps Abroad for Folks at Home. Illustrated throughout.
Wild Adventures in Wild Places. By Dr. Gordon Stables, R.N. Illustrated.
Modern Explorers. By Thomas Frost. Illustrated. *New and Cheaper Edition.*
Early Explorers. By Thomas Frost.
Home Chat with our Young Folks. Illustrated throughout.
Jungle, Peak, and Plain. Illustrated throughout.

---

**The "Cross and Crown" Series.** Illustrated. 2s. 6d. each.

Freedom's Sword: A Story of the Days of Wallace and Bruce. By Annie S. Swan.
Strong to Suffer: A Story of the Jews. By E. Wynne.
Heroes of the Indian Empire; or, Stories of Valour and Victory. By Ernest Foster.
In Letters of Flame: A Story of the Waldenses. By C. L. Matéaux.

Through Trial to Triumph. By Madeline B. Hunt.
By Fire and Sword: A Story of the Huguenots. By Thomas Archer.
Adam Hepburn's Vow: A Tale of Kirk and Covenant. By Annie S. Swan.
No. XIII.; or, The Story of the Lost Vestal. A Tale of Early Christian Days. By Emma Marshall.

---

**"Golden Mottoes" Series, The.** Each Book containing 208 pages, with Four full-page Original Illustrations. Crown 8vo, cloth gilt, 2s. each.

"Nil Desperandum." By the Rev. F. Langbridge, M.A.
"Bear and Forbear." By Sarah Pitt.
"Foremost if I Can." By Helen Atteridge.

"Honour is my Guide." By Jeanie Hering (Mrs. Adams-Acton).
"Aim at a Sure End." By Emily Searchfield.
"He Conquers who Endures." By the Author of "May Cunningham's Trial," &c.

---

**Cassell's Picture Story Books.** Each containing about Sixty Pages of Pictures and Stories, &c. 6d. each.

Little Talks.
Bright Stars.
Nursery Toys.
Pet's Posy.
Tiny Tales.

Daisy's Story Book.
Dot's Story Book.
A Nest of Stories.
Good-Night Stories.
Chats for Small Chatterers.

Auntie's Stories.
Birdie's Story Book.
Little Chimes.
A Sheaf of Tales.
Dewdrop Stories.

---

**Illustrated Books for the Little Ones.** Containing interesting Stories. All Illustrated. 1s. each; cloth gilt, 1s. 6d.

Bright Tales & Funny Pictures.
Merry Little Tales.
Little Tales for Little People.
Little People and Their Pets.
Tales Told for Sunday.
Sunday Stories for Small People.
Stories and Pictures for Sunday.
Bible Pictures for Boys and Girls.
Firelight Stories.
Sunlight and Shade.
Rub-a-Dub Tales.
Fine Feathers and Fluffy Fur.
Scrambles and Scrapes.
Tittle Tattle Tales.

Up and Down the Garden.
All Sorts of Adventures.
Our Sunday Stories.
Our Holiday Hours.
Indoors and Out.
Some Farm Friends.
Wandering Ways.
Dumb Friends.
Those Golden Sands.
Little Mothers & their Children.
Our Pretty Pets.
Our Schoolday Hours.
Creatures Tame.
Creatures Wild.

*Selections from Cassell & Company's Publications.*

**Cassell's Shilling Story Books.** All Illustrated, and containing Interesting Stories.

Bunty and the Boys.
The Heir of Elmdale.
The Mystery at Shoncliff School.
Claimed at Last, & Roy's Reward.
Thorns and Tangles.
The Cuckoo in the Robin's Nest.
John's Mistake. [Pitchers.
The History of Five Little Diamonds in the Sand.
Surly Bob.
The Giant's Cradle.
Shag and Doll.
Aunt Lucia's Locket.
The Magic Mirror.
The Cost of Revenge.
Clever Frank.
Among the Redskins.
The Ferryman of Brill.
Harry Maxwell.
A Banished Monarch.
Seventeen Cats.

**"Wanted—a King"** Series. *Cheap Edition.* Illustrated. 2s. 6d. each.
Great Grandmamma. By Georgina M. Synge.
Robin's Ride. By Ellinor Davenport Adams.
Wanted—a King; or, How Merle set the Nursery Rhymes to Rights. By Maggie Browne. With Original Designs by Harry Furniss.
Fairy Tales in Other Lands. By Julia Goddard.

**The World's Workers.** A Series of New and Original Volumes. With Portraits printed on a tint as Frontispiece. 1s. each.

John Cassell. By G. Holden Pike.
Charles Haddon Spurgeon. By G. Holden Pike.
Dr. Arnold of Rugby. By Rose E. Selfe.
The Earl of Shaftesbury. By Henry Frith.
Sarah Robinson, Agnes Weston, and Mrs. Meredith. By E. M. Tomkinson.
Thomas A. Edison and Samuel F. B. Morse. By Dr. Denslow and J. Marsh Parker.
Mrs. Somerville and Mary Carpenter. By Phyllis Browne.
General Gordon. By the Rev. S. A. Swaine.
Charles Dickens. By his Eldest Daughter.
Sir Titus Salt and George Moore. By J. Burnley.
Florence Nightingale, Catherine Marsh, Frances Ridley Havergal, Mrs. Ranyard ("L. N. R."). By Lizzie Alldridge.
Dr. Guthrie, Father Mathew, Elihu Burritt, George Livesey. By John W. Kirton, LL.D.
Sir Henry Havelock and Colin Campbell Lord Clyde. By E. C. Phillips.
Abraham Lincoln. By Ernest Foster.
George Müller and Andrew Reed. By E. R. Pitman.
Richard Cobden. By R. Gowing.
Benjamin Franklin. By E. M. Tomkinson.
Handel. By Eliza Clarke. [Swaine.
Turner the Artist. By the Rev. S. A.
George and Robert Stephenson. By C. L. Matéaux.
David Livingstone. By Robert Smiles.

\*.\* *The above Works can also be had Three in One Vol., cloth, gilt edges, 3s.*

**Library of Wonders.** Illustrated Gift-books for Boys. Paper, 1s.; cloth, 1s. 6d.

Wonderful Balloon Ascents.
Wonderful Adventures.
Wonderful Escapes.
Wonders of Animal Instinct.
Wonders of Bodily Strength and Skill.

**Cassell's Eighteenpenny Story Books.** Illustrated.

Wee Willie Winkie.
Ups and Downs of a Donkey's Life.
Three Wee Ulster Lassies.
Up the Ladder.
Dick's Hero: and other Stories.
The Chip Boy.
Raggles, Baggles, and the Emperor.
Roses from Thorns.
Faith's Father.
By Land and Sea.
The Young Berringtons.
Jeff and Leff.
Tom Morris's Error.
Worth more than Gold.
"Through Flood—Through Fire;" and other Stories.
The Girl with the Golden Looks.
Stories of the Olden Time.

**Gift Books for Young People.** By Popular Authors. With Four Original Illustrations in each. Cloth gilt, 1s. 6d. each.

The Boy Hunters of Kentucky. By Edward S. Ellis.
Red Feather: a Tale of the American Frontier. By Edward S. Ellis.
Seeking a City.
Rhoda's Reward; or, "If Wishes were Horses."
Jack Marston's Anchor.
Frank's Life-Battle; or, The Three Friends.
Fritters. By Sarah Pitt.
The Two Hardcastles. By Madeline Bonavia Hunt.
Major Monk's Motto. By the Rev. F. Langbridge.
Trixy. By Maggie Symington.
Rags and Rainbows: A Story of Thanksgiving.
Uncle William's Charges; or, The Broken Trust.
Pretty Pink's Purpose; or, The Little Street Merchants.
Tim Thomson's Trial. By George Weatherly.
Ursula's Stumbling-Block. By Julia Goddard.
Ruth's Life-Work. By the Rev. Joseph Johnson.

*Selections from Cassell & Company's Publications.*

### Cassell's Two-Shilling Story Books. Illustrated.

- Margaret's Enemy.
- Stories of the Tower.
- Mr. Burke's Nieces.
- May Cunningham's Trial.
- The Top of the Ladder: How to Reach it.
- Little Flotsam.
- Madge and Her Friends.
- The Children of the Court.
- Maid Marjory.
- Peggy, and other Tales.
- The Four Cats of the Tippertons.
- Marion's Two Homes.
- Little Folks' Sunday Book.
- Two Fourpenny Bits.
- Poor Nelly.
- Tom Heriot.
- Through Peril to Fortune.
- Aunt Tabitha's Waifs.
- In Mischief Again.

### Cheap Editions of Popular Volumes for Young People. Bound in cloth, gilt edges, 2s. 6d. each.

- In Quest of Gold; or, Under the Whanga Falls.
- On Board the *Esmeralda*; or, Martin Leigh's Log.
- For Queen and King.
- Esther West.
- Three Homes.
- Working to Win.
- Perils Afloat and Brigands Ashore.

### Books by Edward S. Ellis. Illustrated. Cloth, 2s. 6d. each.

- The Great Cattle Trail.
- The Path in the Ravine.
- The Young Ranchers.
- The Hunters of the Ozark.
- The Camp in the Mountains.
- Ned in the Woods. A Tale of Early Days in the West.
- Down the Mississippi.
- The Last War Trail.
- Ned on the River. A Tale of Indian River Warfare.
- Footprints in the Forest.
- Up the Tapajos.
- Ned in the Block House. A Story of Pioneer Life in Kentucky.
- The Lost Trail.
- Camp-Fire and Wigwam.
- Lost in the Wilds.
- Lost in Samoa. A Tale of Adventure in the Navigator Islands.
- Tad; or, "Getting Even" with Him.

### The "World in Pictures." Illustrated throughout. *Cheap Edition.* 1s. 6d. each.

- A Ramble Round France.
- All the Russias.
- Chats about Germany.
- The Eastern Wonderland (Japan).
- The Land of Pyramids (Egypt).
- Glimpses of South America.
- Round Africa.
- The Land of Temples (India).
- The Isles of the Pacific.
- Peeps into China

### Half-Crown Story Books.

- Pictures of School Life and Boyhood.
- Pen's Perplexities.
- At the South Pole.

### Books for the Little Ones.

- Rhymes for the Young Folk. By William Allingham. Beautifully Illustrated. 3s. 6d.
- The History Scrap Book. With nearly 1,000 Engravings. Cloth, 7s. 6d.
- The Sunday Scrap Book. With Several Hundred Illustrations. Paper boards, 3s. 6d.; cloth, gilt edges, 5s.
- The Old Fairy Tales. With Original Illustrations. Boards, 1s.; cloth, 1s. 6d.

### Albums for Children. 3s. 6d. each.

- The Album for Home, School, and Play. Containing Stories by Popular Authors. Illustrated.
- My Own Album of Animals. With Full-page Illustrations.
- Picture Album of All Sorts. With Full-page Illustrations.
- The Chit-Chat Album. Illustrated throughout.

**Cassell & Company's Complete Catalogue** *will be sent post free on application to*

CASSELL & COMPANY, Limited, *Ludgate Hill, London.*

www.ingramcontent.com/pod-product-compliance
Lightning Source LLC
Chambersburg PA
CBHW031744230426
43669CB00007B/476